General Theory of Relativity

PRINCETON LANDMARKS IN MATHEMATICS AND PHYSICS

Non-standard Analysis, *by Abraham Robinson*

General Theory of Relativity, *by P.A.M. Dirac*

Angular Momentum in Quantum Mechanics, *by A. R. Edmonds*

Mathematical Foundations of Quantum Mechanics, *by John von Neumann*

Introduction to Mathematical Logic, *by Alonzo Church*

Convex Analysis, *by R. Tyrrell Rockafellar*

Riemannian Geometry, *by Luther Pfahler Eisenhart*

The Classical Groups, *by Hermann Weyl*

Topology from the Differentiable Viewpoint, *by John W. Milnor*

Algebraic Theory of Numbers, *by Hermann Weyl*

Continuous Geometry, *by John von Neumann*

Linear Programming and Extensions, *by George B. Dantzig*

Operator Techniques in Atomic Spectroscopy, *by Brian R. Judd*

The Topology of Fibre Bundles, *by Norman Steenrod*

Mathematical Methods of Statistics, *by Harald Cramér*

GENERAL THEORY OF RELATIVITY

P. A. M. DIRAC

PRINCETON UNIVERSITY PRESS
PRINCETON, NEW JERSEY

Published by Princeton University Press, 41 William Street,
Princeton, New Jersey 08540
In the United Kingdom: Princeton University Press,
Chichester, West Sussex

Library of Congress Cataloging-in-Publication Data
Dirac, P.A.M. (Paul Adrien Maurice), 1902–1984.
General theory of relativity / P.A.M. Dirac.
p. cm. – (Princeton landmarks in mathematics and physics)
(Physics notes)
Originally published: New York : Wiley, [1975].
Includes index.
ISBN 0-691-01146-X (pbk. : alk. paper)
1. General relativity (Physics) I. Title. II. Series.
III. Series: Physics notes.
QC173.6.D57 1996
530.1'1—dc20 95-46196

Princeton University Press books are printed on acid-free paper and meet the guidelines for permanence and durability of the Committee on Production Guidelines for Book Longevity of the Council on Library Resources

First Princeton Paperback printing, in the Princeton Landmarks in Mathematics and Physics Series and the Physics Notes Series, 1996

http://pup.princeton.edu

Printed in the United States of America

5 7 9 10 8 6 4

Preface

Einstein's general theory of relativity requires a curved space for the description of the physical world. If one wishes to go beyond a superficial discussion of the physical relations one needs to set up precise equations for handling curved space. There is a well-established but rather complicated mathematical technique that does this. It has to be mastered by any student who wishes to understand Einstein's theory.

This book is built up from a course of lectures given at the Physics Department of Florida State University and has the aim of presenting the indispensible material in a direct and concise form. It does not require previous knowledge beyond the basic ideas of special relativity and the handling of differentiations of field functions. It will enable the student to pass through the main obstacles of understanding general relativity with the minimum expenditure of time and trouble and to become qualified to continue more deeply into any specialized aspects of the subject that interest him.

P. A. M. Dirac

Tallahassee, Florida
February 1975

Contents

1. Special relativity

For the space-time of physics we need four coordinates, the time t and three space coordinates x, y, z. We put

$$t = x^0, \qquad x = x^1, \qquad y = x^2, \qquad z = x^3,$$

so that the four coordinates may be written x^μ, where the suffix μ takes on the four values 0, 1, 2, 3. The suffix is written in the upper position in order that we may maintain a "balancing" of the suffixes in all the general equations of the theory. The precise meaning of balancing will become clear a little later.

Let us take a point close to the point that we originally considered and let its coordinates be $x^\mu + dx^\mu$. The four quantities dx^μ which form the displacement may be considered as the components of a vector. The laws of special relativity allow us to make linear nonhomogeneous transformations of the coordinates, resulting in linear homogeneous transformations of the dx^μ. These are such that, if we choose units of distance and of time such that the velocity of light is unity,

$$(dx^0)^2 - (dx^1)^2 - (dx^2)^2 - (dx^3)^2 \tag{1.1}$$

is invariant.

Any set of four quantities A^μ that transform under a change of coordinates in the same way as the dx^μ form what is called a *contravariant vector*. The invariant quantity

$$(A^0)^2 - (A^1)^2 - (A^2)^2 - (A^3)^2 = (A, A) \tag{1.2}$$

may be called the squared length of the vector. With a second contravariant vector B^μ, we have the scalar product invariant

$$A^0B^0 - A^1B^1 - A^2B^2 - A^3B^3 = (A, B). \tag{1.3}$$

In order to get a convenient way of writing such invariants we introduce the device of lowering suffixes. Define

$$A_0 = A^0, \qquad A_1 = -A^1, \qquad A_2 = -A^2, \qquad A_3 = -A^3. \tag{1.4}$$

Then the expression on the left-hand side of (1.2) may be written $A_\mu A^\mu$, in which it is understood that a summation is to be taken over the four values of μ. With the same notation we can write (1.3) as $A_\mu B^\mu$ or else $A^\mu B_\mu$.

The four quantities A_μ introduced by (1.4) may also be considered as the components of a vector. Their transformation laws under a change of coordinates are somewhat different from those of the A^μ, because of the differences in sign, and the vector is called a *covariant vector*.

From the two contravariant vectors A^μ and B^μ we may form the sixteen quantities $A^\mu B^\nu$. The suffix ν, like all the Greek suffixes appearing in this work, also takes on the four values 0, 1, 2, 3. These sixteen quantities form the components of a tensor of the second rank. It is sometimes called the outer product of the vectors A^μ and B^μ, as distinct from the scalar product (1.3), which is called the inner product.

The tensor $A^\mu B^\nu$ is a rather special tensor because there are special relations between its components. But we can add together several tensors constructed in this way to get a general tensor of the second rank; say

$$T^{\mu\nu} = A^\mu B^\nu + A'^\mu B'^\nu + A''^\mu B''^\nu + \cdots. \tag{1.5}$$

The important thing about the general tensor is that under a transformation of coordinates its components transform in the same way as the quantities $A^\mu B^\nu$.

We may lower one of the suffixes in $T^{\mu\nu}$ by applying the lowering process to each of the terms on the right-hand side of (1.5). Thus we may form $T_\mu{}^\nu$ or $T^\mu{}_\nu$. We may lower both suffixes to get $T_{\mu\nu}$.

In $T_\mu{}^\nu$ we may set $\nu = \mu$ and get $T_\mu{}^\mu$. This is to be summed over the four values of μ. A summation is always implied over a suffix that occurs twice in a term. Thus $T_\mu{}^\mu$ is a scalar. It is equal to $T^\mu{}_\mu$.

We may continue this process and multiply more than two vectors together, taking care that their suffixes are all different. In this way we can construct tensors of higher rank. If the vectors are all contravariant, we get a tensor with all its suffixes upstairs. We may then lower any of the suffixes and so get a general tensor with any number of suffixes upstairs and any number downstairs.

We may set a downstairs suffix equal to an upstairs one. We then have to sum over all values of this suffix. The suffix becomes a dummy. We are left with a tensor having two fewer effective suffixes than the original one. This process is called *contraction*. Thus, if we start with the fourth-rank tensor $T^\mu{}_{\nu\rho}{}^\sigma$, one way of contracting it is to put $\sigma = \rho$, which gives the second rank tensor $T^\mu{}_{\nu\rho}{}^\rho$, having only sixteen components, arising from the four values of μ and ν. We could contract again to get the scalar $T^\mu{}_{\mu\rho}{}^\rho$, with just one component.

At this stage one can appreciate the balancing of suffixes. Any effective suffix occurring in an equation appears once and only once in each term of the equation, and always upstairs or always downstairs. A suffix occurring twice in a term is a dummy, and it must occur once upstairs and once downstairs. It may be replaced by any other Greek letter not already mentioned in the term. Thus $T^{\mu}{}_{\nu\rho}{}^{\rho} = T^{\mu}{}_{\nu\alpha}{}^{\alpha}$. A suffix must never occur more than twice in a term.

2. Oblique axes

Before passing to the formalism of general relativity it is convenient to consider an intermediate formalism—special relativity referred to oblique rectilinear axes.

If we make a transformation to oblique axes, each of the dx^{μ} mentioned in (1.1) becomes a linear function of the new dx^{μ} and the quadratic form (1.1) becomes a general quadratic form in the new dx^{μ}. We may write it

$$g_{\mu\nu}\, dx^{\mu}\, dx^{\nu}, \tag{2.1}$$

with summations understood over both μ and ν. The coefficients $g_{\mu\nu}$ appearing here depend on the system of oblique axes. Of course we take $g_{\mu\nu} = g_{\nu\mu}$, because any difference of $g_{\mu\nu}$ and $g_{\nu\mu}$ would not show up in the quadratic form (2.1). There are thus ten independent coefficients $g_{\mu\nu}$.

A general contravariant vector has four components A^{μ} which transform like the dx^{μ} under any transformation of the oblique axes. Thus

$$g_{\mu\nu} A^{\mu} A^{\nu}$$

is invariant. It is the squared length of the vector A^{μ}.

Let B^{μ} be a second contravariant vector; then $A^{\mu} + \lambda B^{\mu}$ is still another, for any value of the number λ. Its squared length is

$$g_{\mu\nu}(A^{\mu} + \lambda B^{\mu})(A^{\nu} + \lambda B^{\nu}) = g_{\mu\nu} A^{\mu} A^{\nu} + \lambda(g_{\mu\nu} A^{\mu} B^{\nu} + g_{\mu\nu} A^{\nu} B^{\mu}) + \lambda^2 g_{\mu\nu} B^{\mu} B^{\nu}.$$

This must be an invariant for all values of λ. It follows that the term independent of λ and the coefficients of λ and λ^2 must separately be invariants. The coefficient of λ is

$$g_{\mu\nu}A^{\mu}B^{\nu} + g_{\mu\nu}A^{\nu}B^{\mu} = 2g_{\mu\nu}A^{\mu}B^{\nu},$$

since in the second term on the left we may interchange μ and ν and then use $g_{\mu\nu} = g_{\nu\mu}$. Thus we find that $g_{\mu\nu}A^{\mu}B^{\nu}$ is an invariant. It is the scalar product of A^{μ} and B^{μ}.

Let g be determinant of the $g_{\mu\nu}$. It must not vanish; otherwise the four axes would not provide independent directions in space-time and would not be suitable as axes. For the orthogonal axes of the preceding section the diagonal elements of $g_{\mu\nu}$ are $1, -1, -1, -1$ and the nondiagonal elements are zero. Thus $g = -1$. With oblique axes g must still be negative, because the oblique axes can be obtained from the orthogonal ones by a continuous process, resulting in g varying continuously, and g cannot pass through the value zero.

Define the covariant vector A_{μ}, with a downstairs suffix, by

$$A_{\mu} = g_{\mu\nu}A^{\nu}. \tag{2.2}$$

Since the determinant g does not vanish, these equations can be solved for A^{ν} in terms of the A_{μ}. Let the result be

$$A^{\nu} = g^{\mu\nu}A_{\mu}. \tag{2.3}$$

Each $g^{\mu\nu}$ equals the cofactor of the corresponding $g_{\mu\nu}$ in the determinant of the $g_{\mu\nu}$, divided by the determinant itself. It follows that $g^{\mu\nu} = g^{\nu\mu}$.

Let us substitute for the A^{ν} in (2.2) their values given by (2.3). We must replace the dummy μ in (2.3) by some other Greek letter, say ρ, in order not to have three μ's in the same term. We get

$$A_{\mu} = g_{\mu\nu}g^{\nu\rho}A_{\rho}.$$

Since this equation must hold for any four quantities A_{μ}, we can infer

$$g_{\mu\nu}g^{\nu\rho} = g_{\mu}^{\rho}, \tag{2.4}$$

where

$$\begin{aligned} g_{\mu}^{\rho} &= 1 \quad \text{for } \mu = \rho, \\ &= 0 \quad \text{for } \mu \neq \rho. \end{aligned} \tag{2.5}$$

The formula (2.2) may be used to lower any upper suffix occurring in a tensor. Similarly, (2.3) can be used to raise any downstairs suffix. If a suffix is lowered and raised again, the result is the same as the original tensor, on account of (2.4) and (2.5). Note that g^ρ_μ just produces a substitution of ρ for μ,

$$g^\rho_\mu A^\mu = A^\rho,$$

or of μ for ρ,

$$g^\rho_\mu A_\rho = A_\mu.$$

If we apply the rule for raising a suffix to the μ in $g_{\mu\nu}$, we get

$$g^\alpha{}_\nu = g^{\alpha\mu} g_{\mu\nu}.$$

This agrees with (2.4), if we take into account that in $g^\alpha{}_\nu$ we may write the suffixes one above the other because of the symmetry of $g_{\mu\nu}$. Further we may raise the suffix ν by the same rule and get

$$g^{\alpha\beta} = g^{\nu\beta} g^\alpha_\nu,$$

a result which follows immediately from (2.5). The rules for raising and lowering suffixes apply to all the suffixes in $g_{\mu\nu}$, g^μ_ν, $g^{\mu\nu}$.

3. Curvilinear coordinates

We now pass on to a system of curvilinear coordinates. We shall deal with quantities which are located at a point in space. Such a quantity may have various components, which are then referred to the axes at that point. There may be a quantity of the same nature at all points of space. It then becomes a field quantity.

If we take such a quantity Q (or one of its components if it has several), we can differentiate it with respect to any of the four coordinates. We write the result

$$\frac{\partial Q}{\partial x^\mu} = Q_{,\mu}.$$

A downstairs suffix preceded by a comma will always denote a derivative in this way. We put the suffix μ downstairs in order to balance the upstairs μ

in the denominator on the left. We can see that the suffixes balance by noting that the change in Q, when we pass from the point x^μ to the neighboring point $x^\mu + \delta x^\mu$, is

$$\delta Q = Q_{,\mu}\, \delta x^\mu. \tag{3.1}$$

We shall have vectors and tensors located at a point, with various components referring to the axes at that point. When we change our system of coordinates, the components will change according to the same laws as in the preceding section, depending on the change of axes at the point concerned. We shall have a $g_{\mu\nu}$ and a $g^{\mu\nu}$ to lower and raise suffixes, as before. But *they are no longer constants*. They vary from point to point. They are field quantities.

Let us see the effect of a particular change in the coordinate system. Take new curvilinear coordinates x'^μ, each a function of the four x's. They may be written more conveniently $x^{\mu'}$, with the prime attached to the suffix rather than the main symbol.

Making a small variation in the x^μ, we get the four quantities δx^μ forming the components of a contravariant vector. Referred to the new axes, this vector has the components

$$\delta x^{\mu'} = \frac{\partial x^{\mu'}}{\partial x^\nu}\, \delta x^\nu = x^{\mu'}_{,\nu}\, \delta x^\nu,$$

with the notation of (3.1). This gives the law for the transformation of any contravariant vector A^ν; namely,

$$A^{\mu'} = x^{\mu'}_{,\nu} A^\nu. \tag{3.2}$$

Interchanging the two systems of axes and changing the suffixes, we get

$$A^\lambda = x^\lambda_{,\mu'} A^{\mu'}. \tag{3.3}$$

We know from the laws of partial differentiation that

$$\frac{\partial x^\lambda}{\partial x^{\mu'}} \frac{\partial x^{\mu'}}{\partial x^\nu} = g^\lambda_\nu,$$

with the notation (2.5). Thus

$$x^\lambda_{,\mu'} x^{\mu'}_{,\nu} = g^\lambda_\nu. \tag{3.4}$$

This enables us to see that the two equations (3.2) and (3.3) are consistent, since if we substitute (3.2) into the right-hand side of (3.3), we get

$$x^\lambda_{,\mu'} x^{\mu'}_{,\nu} A^\nu = g^\lambda_\nu A^\nu = A^\lambda.$$

To see how a covariant vector B_μ transforms, we use the condition that $A^\mu B_\mu$ is invariant. Thus with the help of (3.3)

$$A^{\mu'}B_{\mu'} = A^\lambda B_\lambda = x^\lambda_{,\mu'}A^{\mu'}B_\lambda .$$

This result must hold for all values of the four $A^{\mu'}$; therefore we can equate the coefficients of $A^{\mu'}$ and get

$$B_{\mu'} = x^\lambda_{,\mu'}B_\lambda . \tag{3.5}$$

We can now use the formulas (3.2) and (3.5) to transform any tensor with any upstairs and downstairs suffixes. We just have to use coefficients like $x^{\mu'}_{,\nu}$ for each upstairs suffix and like $x^\lambda_{,\mu'}$ for each downstairs suffix and make all the suffixes balance. For example

$$T^{\alpha'\beta'}{}_{\gamma'} = x^{\alpha'}_{,\lambda}x^{\beta'}_{,\mu}x^{\nu}_{,\gamma'}T^{\lambda\mu}{}_{\nu} . \tag{3.6}$$

Any quantity that transforms according to this law is a tensor. This may be taken as the definition of a tensor.

It should be noted that it has a meaning for a tensor to be symmetrical or antisymmetrical between two suffixes like λ and μ, because this property of symmetry is preserved with the change of coordinates.

The formula (3.4) may be written

$$x^\lambda_{,\alpha'}x^{\beta'}_{,\nu}g^{\alpha'}_{\beta'} = g^\lambda_\nu .$$

It just shows that g^λ_ν is a tensor. We have also, for any vectors A^μ, B^ν,

$$g_{\alpha'\beta'}A^{\alpha'}B^{\beta'} = g_{\mu\nu}A^\mu B^\nu = g_{\mu\nu}x^\mu_{,\alpha'}x^\nu_{,\beta'}A^{\alpha'}B^{\beta'} .$$

Since this holds for all values of $A^{\alpha'}$, $B^{\beta'}$, we can infer

$$g_{\alpha'\beta'} = g_{\mu\nu}x^\mu_{,\alpha'}x^\nu_{,\beta'} . \tag{3.7}$$

This shows that $g_{\mu\nu}$ is a tensor. Similarly, $g^{\mu\nu}$ is a tensor. They are called the *fundamental tensors*.

If S is any scalar field quantity, it can be considered either as a function of the four x^μ or of the four $x^{\mu'}$. From the laws of partial differentiation

$$S_{,\mu'} = S_{,\lambda}x^\lambda_{,\mu'} .$$

Hence the $S_{,\lambda}$ transform like the B_λ of equation (3.5) and thus *the derivative of a scalar field is a covariant vector field.*

4. Nontensors

We can have a quantity $N^{\mu}{}_{\nu\rho..}$ with various up and down suffixes, which is not a tensor. If it is a tensor, it must transform under a change of coordinate system according to the law exemplified by (3.6). With any other law it is a nontensor. A tensor has the property that if all the components vanish in one system of coordinates, they vanish in every system of coordinates. This may not hold for a nontensor.

For a nontensor we can raise and lower suffixes by the same rules as for a tensor. Thus, for example,

$$g^{\alpha\nu}N^{\mu}{}_{\nu\rho} = N^{\mu\alpha}{}_{\rho}.$$

The consistency of these rules is quite independent of the transformation laws to a different system of coordinates. Similarly, we can contract a nontensor by putting an upper and lower suffix equal.

We may have tensors and nontensors appearing together in the same equation. The rules for balancing suffixes apply equally to tensors and nontensors.

THE QUOTIENT THEOREM

Suppose $P_{\lambda\mu\nu}$ is such that $A^{\lambda}P_{\lambda\mu\nu}$ is a tensor for any vector A^{λ}. Then $P_{\lambda\mu\nu}$ is a tensor.

To prove it, write $A^{\lambda}P_{\lambda\mu\nu} = Q_{\mu\nu}$. We are given that it is a tensor; therefore

$$Q_{\beta\gamma} = Q_{\mu'\nu'}x^{\mu'}_{,\beta}x^{\nu'}_{,\gamma}.$$

Thus

$$A^{\alpha}P_{\alpha\beta\gamma} = A^{\lambda'}P_{\lambda'\mu'\gamma'}x^{\mu'}_{,\beta}x^{\nu'}_{,\gamma}.$$

Since A^{λ} is a vector, we have from (3.2),

$$A^{\lambda'} = A^{\alpha}x^{\lambda'}_{,\alpha}.$$

So

$$A^{\alpha}P_{\alpha\beta\gamma} = A^{\alpha}x^{\lambda'}_{,\alpha}P_{\lambda'\mu'\nu'}x^{\mu'}_{,\beta}x^{\nu'}_{,\gamma}.$$

This equation must hold for all values of A^α, so

$$P_{\alpha\beta\gamma} = P_{\lambda'\mu'\nu'} x^{\lambda'}_{,\alpha} x^{\mu'}_{,\beta} x^{\nu'}_{,\gamma},$$

showing that $P_{\alpha\beta\gamma}$ is a tensor.

The theorem also holds if $P_{\lambda\mu\nu}$ is replaced by a quantity with any number of suffixes, and if some of the suffixes are upstairs.

5. Curved space

One can easily imagine a curved two-dimensional space as a surface immersed in Euclidean three-dimensional space. In the same way, one can have a curved four-dimensional space immersed in a flat space of a larger number of dimensions. Such a curved space is called a Riemann space. A small region of it is approximately flat.

Einstein assumed that physical space is of this nature and thereby laid the foundation for his theory of gravitation.

For dealing with curved space one cannot introduce a rectilinear system of axes. One has to use curvilinear coordinates, such as those dealt with in Section 3. The whole formalism of that section can be applied to curved space, because all the equations are local ones which are not disturbed by the curvature.

The invariant distance ds between a point x^μ and a neighboring point $x^\mu + dx^\mu$ is given by

$$ds^2 = g_{\mu\nu}\, dx^\mu\, dx^\nu$$

like (2.1). ds is real for a timelike interval and imaginary for a spacelike interval.

With a network of curvilinear coordinates the $g_{\mu\nu}$, given as functions of the coordinates, fix all the elements of distance; so they fix the metric. They determine both the coordinate system and the curvature of the space.

6. Parallel displacement

Suppose we have a vector A^μ located at a point P. If the space is curved, we cannot give a meaning to a parallel vector at a different point Q, as one can easily see if one thinks of the example of a curved two-dimensional space in a three-dimensional Euclidean space. However, if we take a point P' close to P, there is a parallel vector at P', with an uncertainty of the second order, counting the distance from P to P' as the first order. Thus we can give a meaning to displacing the vector A^μ from P to P' keeping it parallel to itself and keeping the length constant.

We can transfer the vector continuously along a path by this process of parallel displacement. Taking a path from P to Q, we end up with a vector at Q which is parallel to the original vector at P with respect to this path. But a different path would give a different result. There is no absolute meaning to a parallel vector at Q. If we transport the vector at P by parallel displacement around a closed loop, we shall end up with a vector at P which is usually in a different direction.

We can get equations for the parallel displacement of a vector by supposing our four-dimensional physical space to be immersed in a flat space of a higher number of dimensions; say N. In this N-dimensional space we introduce rectilinear coordinates $z^n (n = 1, 2, \ldots, N)$. These coordinates do not need to be orthogonal, only rectilinear. Between two neighboring points there is an invariant distance ds given by

$$ds^2 = h_{nm}\, dz^n\, dz^m, \tag{6.1}$$

summed for $n, m = 1, 2, \ldots, N$. The h_{nm} are constants, unlike the $g_{\mu\nu}$. We may use them to lower suffixes in the N-dimensional space; thus

$$dz_n = h_{nm}\, dz^m.$$

Physical space forms a four-dimensional "surface" in the flat N-dimensional space. Each point x^μ in the surface determines a definite point y^n in the N-dimensional space. Each coordinate y^n is a function of the four x's; say $y^n(x)$. The equations of the surface would be given by eliminating the four x's from the $N\, y^n(x)$'s. There are $N - 4$ such equations.

By differentiating the $y^n(x)$ with respect to the parameters x^μ, we get

$$\frac{\partial y^n(x)}{\partial x^\mu} = y^n_{,\mu}.$$

For two neighboring points in the surface differing by δx^μ, we have

$$\delta y^n = y^n_{,\mu}\,\delta x^\mu. \tag{6.2}$$

The squared distance between them is, from (6.1)

$$\delta s^2 = h_{nm}\,\delta y^n\,\delta y^m = h_{nm}\,y^n_{,\mu}\,y^m_{,\nu}\,\delta x^\mu\,\delta x^\nu.$$

We may write it

$$\delta s^2 = y^n_{,\mu}\,y_{n,\nu}\,\delta x^\mu\,\delta x^\nu$$

on account of the h_{nm} being constants. We also have

$$\delta s^2 = g_{\mu\nu}\,\delta x^\mu\,\delta x^\nu.$$

Hence

$$g_{\mu\nu} = y^n_{,\mu}\,y_{n,\nu}. \tag{6.3}$$

Take a contravariant vector A^μ in physical space, located at the point x. Its components A^μ are like the δx^μ of (6.2). They will provide a contravariant vector A^n in the N-dimensional space, like the δy^n of (6.2). Thus

$$A^n = y^n_{,\mu}A^\mu. \tag{6.4}$$

This vector A^n, of course, lies in the surface.

Now shift the vector A^n, keeping it parallel to itself (which means, of course, keeping the components constant), to a neighboring point $x + dx$ in the surface. It will no longer lie in the surface at the new point, on account of the curvature of the surface. But we can project it on to the surface, to get a definite vector lying in the surface.

The projection process consists in splitting the vector into two parts, a tangential part and a normal part, and discarding the normal part. Thus

$$A^n = A^n_{\text{tan}} + A^n_{\text{nor}}. \tag{6.5}$$

Now if K^μ denotes the components of A^n_{tan} referred to the x coordinate system in the surface, we have, corresponding to (6.4),

$$A^n_{\text{tan}} = K^\mu y^n_{,\mu}(x + dx), \tag{6.6}$$

with the coefficients $y^n_{,\mu}$ taken at the new point $x + dx$.

A^n_{nor} is defined to be orthogonal to every tangential vector at the point $x + dx$, and thus to every vector like the right-hand side of (6.6), no matter what the K^μ are. Thus

$$A^n_{\text{nor}}\,y_{n,\mu}(x + dx) = 0.$$

If we now multiply (6.5) by $y_{n,\nu}(x + dx)$, the A^n_{nor} term drops out and we are left with

$$A^n y_{n,\nu}(x + dx) = K^\mu y^n_{,\mu}(x + dx) y_{n,\nu}(x + dx)$$
$$= K^\mu g_{\mu\nu}(x + dx)$$

from (6.3). Thus to the first order in dx

$$K_\nu(x + dx) = A^n[y_{n,\nu}(x) + y_{n,\nu,\sigma}\, dx^\sigma]$$
$$= A^\mu y^n_{,\mu}[y_{n,\nu} + y_{n,\nu,\sigma}\, dx^\sigma]$$
$$= A_\nu + A^\mu y^n_{,\mu} y_{n,\nu,\sigma}\, dx^\sigma.$$

This K_ν is the result of parallel displacement of A_ν to the point $x + dx$. We may put

$$K_\nu - A_\nu = dA_\nu,$$

so dA_ν denotes the change in A_ν under parallel displacement. Then we have

$$dA_\nu = A^\mu y^n_{,\mu} y_{n,\nu,\sigma}\, dx^\sigma. \tag{6.7}$$

7. Christoffel symbols

By differentiating (6.3) we get (omitting the second comma with two differentiations)

$$g_{\mu\nu,\sigma} = y^n_{,\mu\sigma} y_{n,\nu} + y^n_{,\mu} y_{n,\nu\sigma}$$
$$= y_{n,\mu\sigma} y^n_{,\nu} + y_{n,\nu\sigma} y^n_{,\mu}, \tag{7.1}$$

since we can move the suffix n freely up and down, on account of the constancy of the h_{mn}. Interchanging μ and σ in (7.1) we get

$$g_{\sigma\nu,\mu} = y_{n,\sigma\mu} y^n_{,\nu} + y_{n,\nu\mu} y^n_{,\sigma}. \tag{7.2}$$

Interchanging ν and σ in (7.1)

$$g_{\mu\sigma,\nu} = y_{n,\mu\nu} y^n_{,\sigma} + y_{n,\sigma\nu} y^n_{,\mu}. \tag{7.3}$$

Now take (7.1) + (7.3) − (7.2) and divide by 2. The result is

$$\tfrac{1}{2}(g_{\mu\nu,\sigma} + g_{\mu\sigma,\nu} - g_{\nu\sigma,\mu}) = y_{n,\nu\sigma} y^n_{,\mu}. \tag{7.4}$$

Put

$$\Gamma_{\mu\nu\sigma} = \tfrac{1}{2}(g_{\mu\nu,\sigma} + g_{\mu\sigma,\nu} - g_{\nu\sigma,\mu}). \tag{7.5}$$

It is called a Christoffel symbol of the first kind. It is symmetrical between the last two suffixes. It is a nontensor. A simple consequence of (7.5) is

$$\Gamma_{\mu\nu\sigma} + \Gamma_{\nu\mu\sigma} = g_{\mu\nu,\sigma}. \tag{7.6}$$

We see now that (6.7) can be written

$$dA_\nu = A^\mu \Gamma_{\mu\nu\sigma}\, dx^\sigma. \tag{7.7}$$

All reference to the N-dimensional space has now disappeared, as the Christoffel symbol involves only the metric $g_{\mu\nu}$ of physical space.

We can infer that the length of a vector is unchanged by parallel displacement. We have

$$\begin{aligned} d(g^{\mu\nu}A_\mu A_\nu) &= g^{\mu\nu}A_\mu\, dA_\nu + g^{\mu\nu}A_\nu\, dA_\mu + A_\mu A_\nu g^{\mu\nu}{}_{,\sigma}\, dx^\sigma \\ &= A^\nu\, dA_\nu + A^\mu\, dA_\mu + A_\alpha A_\beta g^{\alpha\beta}{}_{,\sigma}\, dx^\sigma \\ &= A^\nu A^\mu \Gamma_{\mu\nu\sigma}\, dx^\sigma + A^\mu A^\nu \Gamma_{\nu\mu\sigma}\, dx^\sigma + A_\alpha A_\beta g^{\alpha\beta}{}_{,\sigma}\, dx^\sigma \\ &= A^\nu A^\mu g_{\mu\nu,\sigma}\, dx^\sigma + A_\alpha A_\beta g^{\alpha\beta}{}_{,\sigma}\, dx^\sigma. \end{aligned} \tag{7.8}$$

Now $g^{\alpha\mu}{}_{,\sigma} g_{\mu\nu} + g^{\alpha\mu} g_{\mu\nu,\sigma} = (g^{\alpha\mu} g_{\mu\nu})_{,\sigma} = g^{\alpha}_{\nu,\sigma} = 0$. Multiplying by $g^{\beta\nu}$, we get

$$g^{\alpha\beta}{}_{,\sigma} = -g^{\alpha\mu} g^{\beta\nu} g_{\mu\nu,\sigma}. \tag{7.9}$$

This is a useful formula giving the derivative of $g^{\alpha\beta}$ in terms of the derivative of $g_{\mu\nu}$. It allows us to infer

$$A_\alpha A_\beta g^{\alpha\beta}{}_{,\sigma} = -A^\mu A^\nu g_{\mu\nu,\sigma}$$

and so the expression (7.8) vanishes. Thus the length of the vector is constant. In particular, a null vector (i.e., a vector of zero length) remains a null vector under parallel displacement.

The constancy of the length of the vector follows also from geometrical arguments. When we split up the vector A^n into tangential and normal parts according to (6.5), the normal part is infinitesimal and is orthogonal to the tangential part. It follows that, to the first order, the length of the whole vector equals that of its tangential part.

The constancy of the length of any vector requires the constancy of the scalar product $g^{\mu\nu}A_\mu B_\nu$ of any two vectors A and B. This can be inferred from the constancy of the length of $A + \lambda B$ for any value of the parameter λ.

It is frequently useful to raise the first suffix of the Christoffel symbol so as to form

$$\Gamma^\mu_{\nu\sigma} = g^{\mu\lambda}\Gamma_{\lambda\nu\sigma}.$$

It is then called a Christoffel symbol of the second kind. It is symmetrical between its two lower suffixes. As explained in Section 4, this raising is quite permissible, even for a nontensor.

The formula (7.7) may be rewritten

$$dA_\nu = \Gamma^\mu_{\nu\sigma} A_\mu \, dx^\sigma. \tag{7.10}$$

It is the standard formula referring to covariant components. For a second vector B^ν we have

$$\begin{aligned} d(A_\nu B^\nu) &= 0 \\ A_\nu \, dB^\nu &= -B^\nu \, dA_\nu = -B^\nu \Gamma^\mu_{\nu\sigma} A_\mu \, dx^\sigma \\ &= -B^\mu \Gamma^\nu_{\mu\sigma} A_\nu \, dx^\sigma. \end{aligned}$$

This must hold for any A_ν, so we get

$$dB^\nu = -\Gamma^\nu_{\mu\sigma} B^\mu \, dx^\sigma. \tag{7.11}$$

This is the standard formula for parallel displacement referring to contravariant components.

8. Geodesics

Take a point with coordinates z^μ and suppose it moves along a track; we then have z^μ a function of some parameter τ. Put $dz^\mu/d\tau = u^\mu$.

There is a vector u^μ at each point of the track. Suppose that as we go along the track the vector u^μ gets shifted by paralled displacement. Then the whole track is determined if we are given the initial point and the initial value of the vector u^μ. We just have to shift the initial point from z^μ to $z^\mu + u^\mu \, d\tau$, then shift the vector u^μ to this new point by parallel displacement, then shift the point again in the direction fixed by the new u^μ, and so on. Not only is the track determined, but also the parameter τ along it. A track produced in this way is called a geodesic.

If the vector u^μ is initially a null vector, it always remains a null vector and the track is called a null geodesic. If the vector u^μ is initially timelike (i.e., $u^\mu u_\mu > 0$), it is always timelike and we have a timelike geodesic. Similarly, if u^μ is initially spacelike ($u^\mu u_\mu < 0$), it is always spacelike and we have a spacelike geodesic.

We get the equations of a geodesic by applying (7.11) with $B^\nu = u^\nu$ and $dx^\sigma = dz^\sigma$. Thus

$$\frac{du^\nu}{d\tau} + \Gamma^\nu_{\mu\sigma} u^\mu \frac{dz^\sigma}{d\tau} = 0 \tag{8.1}$$

or

$$\frac{d^2 z^\nu}{d\tau^2} + \Gamma^\nu_{\mu\sigma} \frac{dz^\mu}{d\tau} \frac{dz^\sigma}{d\tau} = 0. \tag{8.2}$$

For a timelike geodesic we may multiply the initial u^μ by a factor so as to make its length unity. This merely requires a change in the scale of τ. The vector u^μ now always has the length unity. It is just the velocity vector $v^\mu = dz^\mu/ds$, and the parameter τ has become the proper time s.

Equation (8.1) becomes

$$\frac{dv^\mu}{ds} + \Gamma^\mu_{\nu\sigma} v^\nu v^\sigma = 0. \tag{8.3}$$

Equation (8.2) becomes

$$\frac{d^2 z^\mu}{ds^2} + \Gamma^\mu_{\nu\sigma} \frac{dz^\nu}{ds} \frac{dz^\sigma}{ds} = 0. \tag{8.4}$$

We make the physical assumption that the world line of a particle not acted on by any forces, except gravitational, is a timelike geodesic. This replaces Newton's first law of motion. Equation (8.4) fixes the acceleration and provides the equations of motion.

We also make the assumption that the path of a ray of light is a null geodesic. It is fixed by equation (8.2) referring to some parameter τ along the path. The proper time s cannot now be used because ds vanishes.

9. The stationary property of geodesics

A geodesic that is not a null geodesic has the property that $\int ds$, taken along a section of the track with the end points P and Q, is stationary if one makes a small variation of the track keeping the end points fixed.

Let us suppose each point of the track, with coordinates z^μ, is shifted so that its coordinates become $z^\mu + \delta z^\mu$. If dx^μ denotes an element along the track,

$$ds^2 = g_{\mu\nu}\, dx^\mu\, dx^\nu.$$

Thus

$$\begin{aligned} 2\, ds\, \delta(ds) &= dx^\mu\, dx^\nu\, \delta g_{\mu\nu} + g_{\mu\nu}\, dx^\mu\, \delta dx^\nu + g_{\mu\nu}\, dx^\nu\, \delta dx^\mu \\ &= dx^\mu\, dx^\nu g_{\mu\nu,\lambda}\, \delta x^\lambda + 2 g_{\mu\lambda}\, dx^\mu\, \delta dx^\lambda. \end{aligned}$$

Now

$$\delta dx^\lambda = d\delta x^\lambda.$$

Thus, with the help of $dx^\mu = v^\mu\, ds$,

$$\delta(ds) = \left(\tfrac{1}{2} g_{\mu\nu,\lambda} v^\mu v^\nu\, \delta x^\lambda + g_{\mu\lambda} v^\mu \frac{d\delta x^\lambda}{ds}\right) ds.$$

Hence

$$\delta \int ds = \int \delta(ds) = \int \left[\tfrac{1}{2} g_{\mu\nu,\lambda} v^\mu v^\nu\, \delta x^\lambda + g_{\mu\lambda} v^\mu \frac{d\delta x^\lambda}{ds}\right] ds.$$

By partial integration, using the condition that $\delta x^\lambda = 0$ at the end points P and Q, we get

$$\delta \int ds = \int \left[\tfrac{1}{2} g_{\mu\nu,\lambda} v^\mu v^\nu - \frac{d}{ds}(g_{\mu\lambda} v^\mu)\right] \delta x^\lambda\, ds. \tag{9.1}$$

The condition for this to vanish with arbitrary δx^λ is

$$\frac{d}{ds}(g_{\mu\lambda} v^\mu) - \tfrac{1}{2} g_{\mu\nu,\lambda} v^\mu v^\nu = 0. \tag{9.2}$$

Now

$$\begin{aligned} \frac{d}{ds}(g_{\mu\lambda} v^\mu) &= g_{\mu\lambda} \frac{dv^\mu}{ds} + g_{\mu\lambda,\nu} v^\mu v^\nu \\ &= g_{\mu\lambda} \frac{dv^\mu}{ds} + \tfrac{1}{2}(g_{\lambda\mu,\nu} + g_{\lambda\nu,\mu}) v^\mu v^\nu. \end{aligned}$$

Thus the condition (9.2) becomes

$$g_{\mu\lambda}\frac{dv^\mu}{ds} + \Gamma_{\lambda\mu\nu} v^\mu v^\nu = 0.$$

Multiplying this by $g^{\lambda\sigma}$, it becomes

$$\frac{dv^\sigma}{ds} + \Gamma^\sigma_{\mu\nu} v^\mu v^\nu = 0,$$

which is just the condition (8.3) for a geodesic.

This work shows that for a geodesic, (9.1) vanishes and $\int ds$ is stationary. Conversely, if we assume that $\int ds$ is stationary, we can infer that the track is a geodesic. Thus we may use the stationary condition as the definition of a geodesic, except in the case of a null geodesic.

10. Covariant differentiation

Let S be a scalar field. Its derivative $S_{,\nu}$ is a covariant vector, as we saw in Section 3. Now let A_μ be a vector field. Is its derivative $A_{\mu,\nu}$ a tensor?

We must examine how $A_{\mu,\nu}$ transforms under a change of coordinate system. With the notation in Section 3, A_μ transforms to

$$A_{\mu'} = A_\rho x^\rho_{,\mu'}$$

like equation (3.5), and hence

$$\begin{aligned} A_{\mu',\nu'} &= (A_\rho x^\rho_{,\mu'})_{,\nu'} \\ &= A_{\rho,\sigma} x^\sigma_{,\nu'} x^\rho_{,\mu'} + A_\rho x^\rho_{,\mu'\nu'}. \end{aligned}$$

The last term should not be here if we are to have the correct transformation law for a tensor. Thus $A_{\mu,\nu}$ is a nontensor.

We can, however, modify the process of differentiation so as to get a tensor. Let us take the vector A_μ at the point x and shift it to $x + dx$ by parallel displacement. It is still a vector. We may subtract it from the vector A_μ at $x + dx$ and the difference will be a vector. It is, to the first order

$$A_\mu(x + dx) - [A_\mu(x) + \Gamma^\alpha_{\mu\nu} A_\alpha\, dx^\nu] = (A_{\mu,\nu} - \Gamma^\alpha_{\mu\nu} A_\alpha)\, dx^\nu.$$

This quantity is a vector for any vector dx^ν; hence, by the quotient theorem of Section 4, the coefficient

$$A_{\mu,\nu} - \Gamma^\alpha_{\mu\nu} A_\alpha$$

is a tensor. One can easily verify directly that it transforms correctly under a change of coordinate system.

It is called the covariant derivative of A_μ and is written

$$A_{\mu:\nu} = A_{\mu,\nu} - \Gamma^\alpha_{\mu\nu} A_\alpha. \tag{10.1}$$

The sign : before a lower suffix will always denote a covariant derivative, just as the comma denotes an ordinary derivative.

Let B_ν be a second vector. We define the outer product $A_\mu B_\nu$ to have the covariant derivative

$$(A_\mu B_\nu)_{:\sigma} = A_{\mu:\sigma} B_\nu + A_\mu B_{\nu:\sigma}. \tag{10.2}$$

Evidently it is a tensor with three suffixes. It has the value

$$\begin{aligned}(A_\mu B_\nu)_{:\sigma} &= (A_{\mu,\sigma} - \Gamma^\alpha_{\mu\sigma} A_\alpha) B_\nu + A_\mu (B_{\nu,\sigma} - \Gamma^\alpha_{\nu\sigma} B_\alpha) \\ &= (A_\mu B_\nu)_{,\sigma} - \Gamma^\alpha_{\mu\sigma} A_\alpha B_\nu - \Gamma^\alpha_{\nu\sigma} A_\mu B_\alpha.\end{aligned}$$

Let $T_{\mu\nu}$ be a tensor with two suffixes. It is expressible as a sum of terms like $A_\mu B_\nu$, so its covariant derivative is

$$T_{\mu\nu:\sigma} = T_{\mu\nu,\sigma} - \Gamma^\alpha_{\mu\sigma} T_{\alpha\nu} - \Gamma^\alpha_{\nu\sigma} T_{\mu\alpha}. \tag{10.3}$$

The rule can be extended to the covariant derivative of a tensor $Y_{\mu\nu\ldots}$ with any number of suffixes downstairs:

$$Y_{\mu\nu\ldots:\sigma} = Y_{\mu\nu\ldots,\sigma} - \text{a } \Gamma \text{ term for each suffix.} \tag{10.4}$$

In each of these Γ terms we must make the suffixes balance, which is sufficient to fix how the suffixes go.

The case of a scalar is included in the general formula (10.4) with the number of suffixes in Y zero.

$$Y_{:\sigma} = Y_{,\sigma}. \tag{10.5}$$

Let us apply (10.3) to the fundamental tensor $g_{\mu\nu}$. It gives

$$\begin{aligned}g_{\mu\nu:\sigma} &= g_{\mu\nu,\sigma} - \Gamma^\alpha_{\mu\sigma} g_{\alpha\nu} - \Gamma^\alpha_{\nu\sigma} g_{\mu\alpha} \\ &= g_{\mu\nu,\sigma} - \Gamma_{\nu\mu\sigma} - \Gamma_{\mu\nu\sigma} = 0\end{aligned}$$

from (7.6). Thus the $g_{\mu\nu}$ count as constants under covariant differentiation.

Formula (10.2) is the usual rule that one uses for differentiating a product. We assume this usual rule holds also for the covariant derivative of the scalar product of two vectors. Thus

$$(A^{\mu}B_{\mu})_{:\sigma} = A^{\mu}{}_{:\sigma}B_{\mu} + A^{\mu}B_{\mu:\sigma}.$$

We get, according to (10.5) and (10.1),

$$(A^{\mu}B_{\mu})_{,\sigma} = A^{\mu}{}_{:\sigma}B_{\mu} + A^{\mu}(B_{\mu,\sigma} - \Gamma^{\alpha}_{\mu\sigma}B_{\alpha});$$

and hence

$$A^{\mu}_{,\sigma}B_{\mu} = A^{\mu}{}_{:\sigma}B_{\mu} - A^{\alpha}\Gamma^{\mu}_{\alpha\sigma}B_{\mu}.$$

Since this holds for any B_{μ}, we get

$$A^{\mu}{}_{:\sigma} = A^{\mu}_{,\sigma} + \Gamma^{\mu}_{\alpha\sigma}A^{\alpha}, \tag{10.7}$$

which is the basic formula for the covariant derivative of a contravariant vector. The same Christoffel symbol occurs as in the basic formula (10.1) for a covariant vector, but now there is a + sign. The arrangement of the suffixes is completely determined by the balancing requirement.

We can extend the formalism so as to include the covariant derivative of any tensor with any number of upstairs and downstairs suffiex. A Γ term appears for each suffix, with a + sign if the suffix is upstairs and a − sign if it is downstairs. If we contract two suffixes in the tensor, the corresponding Γ terms cancel.

The formula for the covariant derivative of a product,

$$(XY)_{:\sigma} = X_{:\sigma}Y + XY_{:\sigma}, \tag{10.8}$$

holds quite generally, with X and Y any kind of tensor quantities. On account of the $g_{\mu\nu}$ counting as constants, we can shift suffixes up or down before covariant differentiation and the result is the same as if we shifted them afterwards.

The covariant derivative of a nontensor has no meaning.

The laws of physics must be valid in all systems of coordinates. They must thus be expressible as tensor equations. Whenever they involve the derivative of a field quantity, it must be a covariant derivative. The field equations of physics must all be rewritten with the ordinary derivatives replaced by covariant derivatives. For example, the d'Alembert equation $\Box V = 0$ for a scalar V becomes, in covariant form

$$g^{\mu\nu}V_{:\mu:\nu} = 0.$$

This gives, from (10.1) and (10.5),

$$g^{\mu\nu}(V_{,\mu\nu} - \Gamma^{\alpha}_{\mu\nu} V_{,\alpha}) = 0. \tag{10.9}$$

Even if one is working with flat space (which means neglecting the gravitational field) and one is using curvilinear coordinates, one must write one's equations in terms of covariant derivatives if one wants them to hold in all systems of coordinates.

11. The curvature tensor

With the product law (10.8) we see that covariant differentiation is very similar to ordinary differentiations. But there is an important property of ordinary differentiation, that if we perform two differentiations in succession their order does not matter, which does not, in general, hold for covariant differentiation.

Let us first consider a scalar field S. We have from the formula (10.1),

$$\begin{aligned} S_{:\mu:\nu} &= S_{:\mu,\nu} - \Gamma^{\alpha}_{\mu\nu} S_{:\alpha} \\ &= S_{,\mu\nu} - \Gamma^{\alpha}_{\mu\nu} S_{,\alpha}. \end{aligned} \tag{11.1}$$

This is symmetrical between μ and ν, so in this case the order of the covariant differentiations does not matter.

Now let us take a vector A_ν and apply two covariant differentiations to it. From the formula (10.3) with $A_{\nu:\rho}$ for $T_{\nu\rho}$ we get

$$\begin{aligned} A_{\nu:\rho:\sigma} &= A_{\nu:\rho,\sigma} - \Gamma^{\alpha}_{\nu\sigma} A_{\alpha:\rho} - \Gamma^{\alpha}_{\rho\sigma} A_{\nu:\alpha} \\ &= (A_{\nu,\rho} - \Gamma^{\alpha}_{\nu\rho} A_\alpha)_{,\sigma} - \Gamma^{\alpha}_{\nu\sigma}(A_{\alpha,\rho} - \Gamma^{\beta}_{\alpha\rho} A_\beta) - \Gamma^{\alpha}_{\rho\sigma}(A_{\nu,\alpha} - \Gamma^{\beta}_{\nu\alpha} A_\beta) \\ &= A_{\nu,\rho,\sigma} - \Gamma^{\alpha}_{\nu\rho} A_{\alpha,\sigma} - \Gamma^{\alpha}_{\nu\sigma} A_{\alpha,\rho} - \Gamma^{\alpha}_{\rho\sigma} A_{\nu,\alpha} \\ &\quad - A_\beta(\Gamma^{\beta}_{\nu\rho,\sigma} - \Gamma^{\alpha}_{\nu\sigma}\Gamma^{\beta}_{\alpha\rho} - \Gamma^{\alpha}_{\rho\sigma}\Gamma^{\beta}_{\nu\alpha}). \end{aligned}$$

Interchange ρ and σ here and subtract from the previous expression. The result is

$$A_{\nu:\rho:\sigma} - A_{\nu:\sigma:\rho} = A_\beta R^{\beta}_{\nu\rho\sigma}, \tag{11.2}$$

where

$$R^{\beta}_{\nu\rho\sigma} = \Gamma^{\beta}_{\nu\sigma,\rho} - \Gamma^{\beta}_{\nu\rho,\sigma} + \Gamma^{\alpha}_{\nu\sigma}\Gamma^{\beta}_{\alpha\rho} - \Gamma^{\alpha}_{\nu\rho}\Gamma^{\beta}_{\alpha\sigma}. \tag{11.3}$$

The left-hand side of (11.2) is a tensor. It follows that the right-hand side of (11.2) is a tensor. This holds for any vector A_β; therefore, by the quotient theorem in Section 4, $R^{\beta}_{\nu\rho\sigma}$ is a tensor. It is called the Riemann-Christoffel tensor or the curvature tensor.

It has the obvious property

$$R^{\beta}_{\nu\rho\sigma} = -R^{\beta}_{\nu\sigma\rho}. \tag{11.4}$$

Also, we easily see from (11.3) that

$$R^{\beta}_{\nu\rho\sigma} + R^{\beta}_{\rho\sigma\nu} + R^{\beta}_{\sigma\nu\rho} = 0. \tag{11.5}$$

Let us lower the suffix β and put it as the first suffix. We get

$$R_{\mu\nu\rho\sigma} = g_{\mu\beta}R^{\beta}_{\nu\rho\sigma} = g_{\mu\beta}\Gamma^{\beta}_{\nu\sigma,\rho} + \Gamma^{\alpha}_{\nu\sigma}\Gamma_{\mu\alpha\rho} - \langle\rho\sigma\rangle,$$

where the symbol $\langle\rho\sigma\rangle$ is used to denote the preceding terms with ρ and σ interchanged. Thus

$$\begin{aligned} R_{\mu\nu\rho\sigma} &= \Gamma_{\mu\nu\sigma,\rho} - g_{\mu\beta,\rho}\Gamma^{\beta}_{\nu\sigma} + \Gamma_{\mu\beta\rho}\Gamma^{\beta}_{\nu\sigma} - \langle\rho\sigma\rangle \\ &= \Gamma_{\mu\nu\sigma,\rho} - \Gamma_{\beta\mu\rho}\Gamma^{\beta}_{\nu\sigma} - \langle\rho\sigma\rangle, \end{aligned}$$

from (7.6). So from (7.5)

$$R_{\mu\nu\rho\sigma} = \tfrac{1}{2}(g_{\mu\sigma,\nu\rho} - g_{\nu\sigma,\mu\rho} - g_{\mu\rho,\nu\sigma} + g_{\nu\rho,\mu\sigma}) + \Gamma_{\beta\mu\sigma}\Gamma^{\beta}_{\nu\rho} - \Gamma_{\beta\mu\rho}\Gamma^{\beta}_{\nu\sigma}. \tag{11.6}$$

Some further symmetries now show up; namely,

$$R_{\mu\nu\rho\sigma} = -R_{\nu\mu\rho\sigma} \tag{11.7}$$

and

$$R_{\mu\nu\rho\sigma} = R_{\rho\sigma\mu\nu} = R_{\sigma\rho\nu\mu}. \tag{11.8}$$

The result of all these symmetries is that, of the 256 components of $R_{\mu\nu\rho\sigma}$, only 20 are independent.

12. The condition for flat space

If space is flat, we may choose a system of coordinates that is rectilinear, and then the $g_{\mu\nu}$ are constant. The tensor $R_{\mu\nu\rho\sigma}$ then vanishes.

Conversely, if $R_{\mu\nu\rho\sigma}$ vanishes, one can prove that the space is flat. Let us take a vector A_μ situated at the point x and shift it by parallel displacement to the point $x + dx$. Then shift it by parallel displacement to the point $x + dx + \delta x$. If $R_{\mu\nu\rho\sigma}$ vanishes, the result must be the same as if we had shifted it first from x to $x + \delta x$, then to $x + \delta x + dx$. Thus we can shift the vector to a distant point and the result we get is independent of the path to the distant point. Therefore, if we shift the original vector A_μ at x to all points by parallel displacement, we get a vector field that satisfies $A_{\mu:\nu} = 0$, or

$$A_{\mu,\nu} = \Gamma^\sigma_{\mu\nu} A_\sigma. \tag{12.1}$$

Can such a vector field be the gradient of a scalar? Let us put $A_\mu = S_{,\mu}$ in (12.1). We get

$$S_{,\mu\nu} = \Gamma^\sigma_{\mu\nu} S_{,\sigma}. \tag{12.2}$$

On account of the symmetry of $\Gamma^\sigma_{\mu\nu}$ in the lower suffixes, we have the same value for $S_{,\mu\nu}$ as $S_{,\nu\mu}$ and the equations (12.2) are integrable.

Let us take four independent scalars satisfying (12.2) and let us take them to be the coordinates $x^{\alpha'}$ of a new system of coordinates. Then

$$x^{\alpha'}_{,\mu\nu} = \Gamma^\sigma_{\mu\nu} x^{\alpha'}_{,\sigma}.$$

According to the transformation law (3.7),

$$g_{\mu\lambda} = g_{\alpha'\beta'} x^{\alpha'}_{,\mu} x^{\beta'}_{,\lambda}.$$

Differentiating this equation with respect to x^ν, we get

$$\begin{aligned} g_{\mu\lambda,\nu} - g_{\alpha'\beta',\nu} x^{\alpha'}_{,\mu} x^{\beta'}_{,\lambda} &= g_{\alpha'\beta'}(x^{\alpha'}_{,\mu\nu} x^{\beta'}_{,\lambda} + x^{\alpha'}_{,\mu} x^{\beta'}_{,\lambda\nu}) \\ &= g_{\alpha'\beta'}(\Gamma^\sigma_{\mu\nu} x^{\alpha'}_{,\sigma} x^{\beta'}_{,\lambda} + x^{\alpha'}_{,\mu} \Gamma^\sigma_{\lambda\nu} x^{\beta'}_{,\sigma}) \\ &= g_{\sigma\lambda} \Gamma^\sigma_{\mu\nu} + g_{\mu\sigma} \Gamma^\sigma_{\lambda\nu} \\ &= \Gamma_{\lambda\mu\nu} + \Gamma_{\mu\lambda\nu} = g_{\mu\lambda,\nu} \end{aligned}$$

from (7.6). Thus

$$g_{\alpha'\beta',\nu} x^{\alpha'}_{,\mu} x^{\beta'}_{,\lambda} = 0.$$

It follows that $g_{\alpha'\beta',\nu} = 0$. Referred to the new system of coordinates, the fundamental tensor is constant. Thus we have flat space referred to rectilinear coordinates.

13. The Bianci relations

To deal with the second covariant derivative of a tensor, take first the case in which the tensor is the outer product of two vectors $A_\mu B_\tau$. We have

$$\begin{aligned}(A_\mu B_\tau)_{:\rho:\sigma} &= (A_{\mu:\rho} B_\tau + A_\mu B_{\tau:\rho})_{:\sigma} \\ &= A_{\mu:\rho:\sigma} B_\tau + A_{\mu:\rho} B_{\tau:\sigma} + A_{\mu:\sigma} B_{\tau:\rho} + A_\mu B_{\tau:\rho:\sigma}.\end{aligned}$$

Now interchange ρ and σ and subtract. We get from (11.2)

$$(A_\mu B_\tau)_{:\rho:\sigma} - (A_\mu B_\tau)_{:\sigma:\rho} = A_\alpha R^\alpha_{\mu\rho\sigma} B_\tau + A_\mu R^\alpha_{\tau\rho\sigma} B_\alpha.$$

A general tensor $T_{\mu\tau}$ is expressible as a sum of terms like $A_\mu B_\tau$, so it must satisfy

$$T_{\mu\tau:\rho:\sigma} - T_{\mu\tau:\sigma:\rho} = T_{\alpha\tau} R^\alpha_{\mu\rho\sigma} + T_{\mu\alpha} R^\alpha_{\tau\rho\sigma}. \tag{13.1}$$

Now take $T_{\mu\tau}$ to be the covariant derivative of a vector $A_{\mu:\tau}$. We get

$$A_{\mu:\tau:\rho:\sigma} - A_{\mu:\tau:\sigma:\rho} = A_{\alpha:\tau} R^\alpha_{\mu\rho\sigma} + A_{\mu:\alpha} R^\alpha_{\tau\rho\sigma}.$$

In this formula make cyclic permutations of τ, ρ, σ and add the three equations so obtained. The left-hand side gives

$$\begin{aligned}A_{\mu:\rho:\sigma:\tau} &- A_{\mu:\sigma:\rho:\tau} + \text{cyc perm} \\ &= (A_\alpha R^\alpha_{\mu\rho\sigma})_{:\tau} + \text{cyc perm} \\ &= A_{\alpha:\tau} R^\alpha_{\mu\rho\sigma} + A_\alpha R^\alpha_{\mu\rho\sigma:\tau} + \text{cyc perm}.\end{aligned} \tag{13.2}$$

The right-hand side gives

$$A_{\alpha:\tau} R^\alpha_{\mu\rho\sigma} + \text{cyc perm}, \tag{13.3}$$

as the remaining terms cancel from (11.5). The first term of (13.2) cancels with (13.3) and we are left with

$$A_\alpha R^\alpha_{\mu\rho\sigma:\tau} + \text{cyc perm} = 0.$$

The factor A_α occurs throughout this equation and may be canceled out. We are left with

$$R^\alpha_{\mu\rho\sigma:\tau} + R^\alpha_{\mu\sigma\tau:\rho} + R^\alpha_{\mu\tau\rho:\sigma} = 0. \tag{13.4}$$

The curvature tensor satisfies these differential equations as well as all the symmetry relations in Section 11. They are known as the Bianci relations.

14. The Ricci tensor

Let us contract two of the suffixes in $R_{\mu\nu\rho\sigma}$. If we take two with respect to which it is antisymmetrical, we get zero, of course. If we take any other two we get the same result, apart from the sign, because of the symmetries (11.4), (11.7), and (11.8). Let us take the first and last and put

$$R^{\mu}_{\nu\rho\mu} = R_{\nu\rho}.$$

It is called the Ricci tensor.

By multiplying (11.8) by $g^{\mu\sigma}$ we get

$$R_{\nu\rho} = R_{\rho\nu}. \tag{14.1}$$

The Ricci tensor is symmetrical.

We may contract again and form

$$g^{\nu\rho}R_{\nu\rho} = R^{\nu}_{\nu} = R,$$

say. This R is a scalar and is called the scalar curvature or total curvature. It is defined in such a way that it is positive for the surface of a sphere in three dimensions, as one can check by a straightforward calculation.

The Bianci relation (13.4) involves five suffixes. Let us contract it twice and get a relation with one nondummy suffix. Put $\tau = \alpha$ and multiply by $g^{\mu\rho}$. The result is

$$g^{\mu\rho}(R^{\alpha}_{\mu\rho\sigma:\alpha} + R^{\alpha}_{\mu\sigma\alpha:\rho} + R^{\alpha}_{\mu\alpha\rho:\sigma}) = 0$$

or

$$(g^{\mu\rho}R^{\alpha}_{\mu\rho\sigma})_{:\alpha} + (g^{\mu\rho}R^{\alpha}_{\mu\sigma\alpha})_{:\rho} + (g^{\mu\rho}R^{\alpha}_{\mu\alpha\rho})_{:\sigma} = 0. \tag{14.2}$$

Now

$$\begin{aligned} g^{\mu\rho}R^{\alpha}_{\mu\rho\sigma} &= g^{\mu\rho}g^{\alpha\beta}R_{\beta\mu\rho\sigma} = g^{\mu\rho}g^{\alpha\beta}R_{\mu\beta\sigma\rho} \\ &= g^{\alpha\beta}R_{\beta\sigma} = R^{\alpha}_{\sigma}. \end{aligned}$$

One can write R^{α}_{σ} with the suffixes one over the other on account of $R_{\alpha\sigma}$ being symmetrical. Equation (14.2) now becomes

$$R^{\alpha}_{\sigma:\alpha} + (g^{\mu\rho}R_{\mu\sigma})_{:\rho} - R_{:\sigma} = 0$$

or

$$2R^{\alpha}_{\sigma:\alpha} - R_{:\sigma} = 0,$$

which is the Bianci relation for the Ricci tensor. If we raise the suffix σ, we get

$$(R^{\sigma\alpha} - \tfrac{1}{2}g^{\sigma\alpha}R)_{:\alpha} = 0. \tag{14.3}$$

The explicit expression for the Ricci tensor is, from (11.3)

$$R_{\mu\nu} = \Gamma^{\alpha}_{\mu\alpha,\nu} - \Gamma^{\alpha}_{\mu\nu,\alpha} - \Gamma^{\alpha}_{\mu\nu}\Gamma^{\beta}_{\alpha\beta} + \Gamma^{\alpha}_{\mu\beta}\Gamma^{\beta}_{\nu\alpha}. \tag{14.4}$$

The first term here does not appear to be symmetrical in μ and ν, although the other three terms evidently are. To establish that the first term really is symmetrical we need a little calculation.

To differentiate the determinant g we must differentiate each element $g_{\lambda\mu}$ in it and then multiply by the cofactor $gg^{\lambda\mu}$. Thus

$$g_{,\nu} = gg^{\lambda\mu}g_{\lambda\mu,\nu}. \tag{14.5}$$

Hence

$$\begin{aligned}\Gamma^{\mu}_{\nu\mu} &= g^{\lambda\mu}\Gamma_{\lambda\nu\mu} = \tfrac{1}{2}g^{\lambda\mu}(g_{\lambda\nu,\mu} + g_{\lambda\mu,\nu} - g_{\mu\nu,\lambda}) \\ &= \tfrac{1}{2}g^{\lambda\mu}g_{\lambda\mu,\nu} = \tfrac{1}{2}g^{-1}g_{,\nu} = \tfrac{1}{2}(\log g)_{,\nu}.\end{aligned} \tag{14.6}$$

This makes it evident that the first term of (14.4) is symmetrical.

15. Einstein's law of gravitation

Up to the present our work has all been pure mathematics (apart from the physical assumption that the track of a particle is a geodesic). It was done mainly in the last century and applies to curved space in any number of dimensions. The only place where the number of dimensions would appear in the formalism is in the equation

$$g^{\mu}_{\mu} = \text{number of dimensions.}$$

Einstein made the assumption that in empty space

$$R_{\mu\nu} = 0. \tag{15.1}$$

It constitutes his law of gravitation. "Empty" here means that there is no matter present and no physical fields except the gravitational field. The gravitational field does not disturb the emptyness. Other fields do. The conditions for empty space hold in a good approximation for the space between the planets in the solar system and equation (15.1) applies there.

Flat space obviously satisfies (15.1). The geodesics are then straight lines and so particles move along straight lines. Where space is not flat, Einstein's law puts restrictions on the curvature. Combined with the assumption that the planets move along geodesics, it gives some information about their motion.

At first sight Einstein's law of gravitation does not look anything like Newton's. To see a similarity, we must look on the $g_{\mu\nu}$ as *potentials* describing the gravitational field. There are ten of them, instead of just the one potential of the Newtonian theory. They describe not only the gravitational field, but also the system of coordinates. The gravitational field and the system of coordinates are inextricably mixed up in the Einstein theory, and one cannot describe the one without the other.

Looking upon the $g_{\mu\nu}$ as potentials, we find that (15.1) appears as field equations. They are like the usual field equations of physics in that they are of the second order, because second derivatives appear in (14.4), as the Christoffel symbols involve first derivatives. They are unlike the usual field equations in that they are not linear; far from it. The nonlinearity means that the equations are complicated and it is difficult to get accurate solutions.

16. The Newtonian approximation

Let us consider a static gravitational field and refer it to a static coordinate system. The $g_{\mu\nu}$ are then constant in time, $g_{\mu\nu,0} = 0$. Further, we must have

$$g_{m0} = 0, \qquad (m = 1, 2, 3).$$

This leads to

$$g^{m0} = 0, \qquad g^{00} = (g_{00})^{-1},$$

and g^{mn} is the reciprocal matrix to g_{mn}. Roman suffixes like m and n always take on the values 1, 2, 3. We find that $\Gamma_{m0n} = 0$, and hence also $\Gamma^m_{0n} = 0$.

Let us take a particle that is moving slowly, compared with the velocity of light. Then v^m is a small quantity, of the first order. With neglect of second-order quantities,

$$g_{00}v^{0^2} = 1. \tag{16.1}$$

The particle will move along a geodesic. With neglect of second-order quantities, the equation (8.3) gives

$$\frac{dv^m}{ds} = -\Gamma^m_{00}v^{0^2} = -g^{mn}\Gamma_{n00}v^{0^2}$$

$$= \tfrac{1}{2}g^{mn}g_{00,n}v^{0^2}.$$

Now

$$\frac{dv^m}{ds} = \frac{dv^m}{dx^\mu}\frac{dx^\mu}{ds} = \frac{dv^m}{dx^0}v^0$$

to the first order. Thus

$$\frac{dv^m}{dx^0} = \tfrac{1}{2}g^{mn}g_{00,n}v^0 = g^{mn}(g_{00}{}^{1/2})_{,n} \tag{16.2}$$

with the help of (16.1). Since the $g_{\mu\nu}$ are independent of x^0, we may lower the suffix m here and get

$$\frac{dv_m}{dx^0} = (g_{00}{}^{1/2})_{,m}. \tag{16.3}$$

We see that the particle moves as though it were under the influence of a potential $g_{00}{}^{1/2}$. We have not used Einstein's law to obtain this result. We now use Einstein's law to obtain a condition for the potential, so that it can be compared with Newton's.

Let us suppose that the gravitational field is weak, so that the curvature of space is small. Then we may choose our coordinate system so that the curvature of the coordinate lines (each with three x's constant) is small. Under these conditions the $g_{\mu\nu}$ are approximately constant, and $g_{\mu\nu,\sigma}$ and all the Christoffel symbols are small. If we count them of the first order and neglect second-order quantities, Einstein's law (15.1) becomes, from (14.4)

$$\Gamma^\alpha_{\mu\alpha,\nu} - \Gamma^\alpha_{\mu\nu,\alpha} = 0.$$

We can evaluate this most conveniently by contracting (11.6) with ρ and μ interchanged and neglecting second-order terms. The result is

$$g^{\rho\sigma}(g_{\rho\sigma,\mu\nu} - g_{\nu\sigma,\mu\rho} - g_{\mu\rho,\nu\sigma} + g_{\mu\nu,\rho\sigma}) = 0. \tag{16.4}$$

Now take $\mu = \nu = 0$ and use the condition that the $g_{\mu\nu}$ are independent of x^0. We get

$$g^{mn} g_{00,mn} = 0. \quad (16.5)$$

The d'Alembert equation (10.9) becomes, in the weak field approximation,

$$g^{\mu\nu} V_{,\mu\nu} = 0.$$

In the static case this reduces to the Laplace equation

$$g^{mn} V_{,mn} = 0.$$

Equation (16.5) just tells us that g_{00} satisfies the Laplace equation.

We may choose our unit of time so that g_{00} is approximately unity. Then we may put

$$g_{00} = 1 + 2V, \quad (16.6)$$

with V small. We get ${g_{00}}^{1/2} = 1 + V$ and V becomes the potential. It satisfies the Laplace equation, so that it can be identified with the Newtonian potential, equal to $-m/r$ for a mass m at the origin. To check the sign we see that (16.2) leads to

$$\text{acceleration} = -\text{grad } V,$$

since g^{mn} has the diagonal elements approximately -1.

We see that Einstein's law of gravitation goes over to Newton's when the field is weak and when it is static. The successes of the Newtonian theory in explaining the motions of the planets can thus be preserved. The static approximation is a good one because the velocities of the planets are all small compared with the velocity of light. The weak field approximation is a good one because the space is very nearly flat. Let us consider some orders of magnitude.

The value of $2V$ on the surface of the earth turns out to be of the order 10^{-9}. Thus g_{00} given by (16.6) is very close to 1. Even so, its difference from 1 is big enough to produce the important gravitational effects that we see on earth. Taking the earth's radius to be of the order 10^9 cm, we find that $g_{00,m}$ is of the order 10^{-18} cm^{-1}. The departure from flatness is thus extremely small. However, this has to be multiplied by the square of the velocity of light, namely 9×10^{20} (cm/sec)^2, to give the acceleration due to gravity at the earth's surface. Thus this acceleration, about 10^3 cm/sec^2, is quite appreciable, even though the departure from flatness is far too small to be observed directly.

17. The gravitational red shift

Let us take again a static gravitational field and consider an atom at rest emitting monochromatic radiation. The wavelength of the light will correspond to a definite Δs. Since the atom is at rest we have, for a static system of coordinates such as we used in Section 16,

$$\Delta s^2 = g_{00} \Delta x^{0^2},$$

where Δx^0 is the period, that is, the time between successive crests referred to our static coordinate system.

If the light travels to another place, Δx^0 will remain constant. This Δx^0 will not be the same as the period of the same spectral line emitted by a local atom, which would be Δs again. The period is thus dependent on the gravitational potential g_{00} at the place where the light was emitted:

$$\Delta x^0 :: {g_{00}}^{-1/2}.$$

The spectral line will be shifted by this factor ${g_{00}}^{-1/2}$.

If we use the Newtonian approximation (16.6), we have

$$\Delta x^0 :: 1 - V.$$

V will be negative at a place with a strong gravitational field, such as the surface of the sun, so light emitted there will be red-shifted when compared with the corresponding light emitted on earth. The effect can be observed with the sun's light but is rather masked by other physical effects, such as the Doppler effect arising from the motion of the emitting atoms. It can be better observed in light emitted from a white dwarf star, where the high density of the matter in the star gives rise to a much stronger gravitational potential at its surface.

18. The Schwarzschild solution

The Einstein equations for empty space are nonlinear and are therefore very complicated, and it is difficult to get accurate solutions of them. There is, however, one special case which can be solved without too much trouble; namely, the static spherically symmetric field produced by a spherically symmetric body at rest.

The static condition means that, with a static coordinate system, the $g_{\mu\nu}$ are independent of the time x^0 or t and also $g_{0m} = 0$. The spatial coordinates may be taken to be spherical polar coordinates $x^1 = r$, $x^2 = \theta$, $x^3 = \phi$. The most general form for ds^2 compatible with spherical symmetry is

$$ds^2 = U\,dt^2 - V\,dr^2 - Wr^2(d\theta^2 + \sin^2\theta\,d\phi^2),$$

where U, V, and W are functions of r only. We may replace r by any function of r without disturbing the spherical symmetry. We use this freedom to simplify things as much as possible, and the most convenient arrangement is to have $W = 1$. The expression for ds^2 may then be written

$$ds^2 = e^{2\nu}\,dt^2 - e^{2\lambda}\,dr^2 - r^2\,d\theta^2 - r^2\sin^2\theta\,d\phi^2, \qquad (18.1)$$

with ν and λ functions of r only. They must be chosen to satisfy the Einstein equations.

We can read off the values of the $g_{\mu\nu}$ from (18.1), namely,

$$g_{00} = e^{2\nu}, \qquad g_{11} = -e^{2\lambda}, \qquad g_{22} = -r^2, \qquad g_{33} = -r^2\sin^2\theta,$$

and

$$g_{\mu\nu} = 0 \quad \text{for} \quad \mu \neq \nu.$$

We find

$$g^{00} = e^{-2\nu}, \qquad g^{11} = -e^{-2\lambda}, \qquad g^{22} = -r^{-2}, \qquad g^{33} = -r^{-2}\sin^{-2}\theta,$$

and

$$g^{\mu\nu} = 0 \quad \text{for} \quad \mu \neq \nu.$$

It is now necessary to calculate all the Christoffel symbols $\Gamma^\sigma_{\mu\nu}$. Many of them vanish. The ones that do not are, with primes denoting differentiations with respect to r,

$$\Gamma^1_{00} = \nu' e^{2\nu-2\lambda} \qquad \Gamma^0_{10} = \nu'$$
$$\Gamma^1_{11} = \lambda' \qquad \Gamma^2_{12} = \Gamma^3_{13} = r^{-1}$$
$$\Gamma^1_{22} = -re^{-2\lambda} \qquad \Gamma^3_{23} = \cot\theta$$
$$\Gamma^1_{33} = -r\sin^2\theta\, e^{-2\lambda} \qquad \Gamma^2_{33} = -\sin\theta\cos\theta.$$

These expressions are to be substituted in (14.4). The results are

$$R_{00} = \left(-\nu'' + \lambda'\nu' - \nu'^2 - \frac{2\nu'}{r}\right)e^{2\nu-2\lambda}, \tag{18.2}$$

$$R_{11} = \nu'' - \lambda'\nu' + \nu'^2 - \frac{2\lambda'}{r} \tag{18.3}$$

$$R_{22} = (1 + r\nu' - r\lambda')e^{-2\lambda} - 1 \tag{18.4}$$

$$R_{33} = R_{22}\sin^2\theta,$$

with the other components of $R_{\mu\nu}$ vanishing.

Einstein's law of gravitation requires these expressions to vanish. The vanishing of (18.2) and (18.3) leads to

$$\lambda' + \nu' = 0.$$

For large values of r the space must approximate to being flat, so that λ and ν both tend to zero as $r \to \infty$. It follows that

$$\lambda + \nu = 0.$$

The vanishing of (18.4) now gives

$$(1 + 2r\nu')e^{2\nu} = 1$$

or

$$(re^{2\nu})' = 1.$$

Thus

$$re^{2\nu} = r - 2m,$$

where m is a constant of integration. This also makes (18.2) and (18.3) vanish. We now get

$$g_{00} = 1 - \frac{2m}{r}. \tag{18.5}$$

The Newtonian approximation must hold for large values of r. Comparing (18.5) with (16.6), we see that the constant of integration m that has appeared in (18.5) is just the mass of the central body that is producing the gravitational field.

The complete solution is

$$ds^2 = \left(1 - \frac{2m}{r}\right) dt^2 - \left(1 - \frac{2m}{r}\right)^{-1} dr^2 - r^2\, d\theta^2 - r^2 \sin^2\theta\, d\phi^2. \qquad (18.6)$$

It is known as the Schwarzschild solution. It holds outside the surface of the body that is producing the field, where there is no matter. Thus it holds fairly accurately outside the surface of a star.

The solution (18.6) leads to small corrections in the Newtonian theory for the motions of the planets around the Sun. These corrections are appreciable only in the case of Mercury, the nearest planet, and they explain the discrepancy of the motion of this planet with the Newtonian theory. Thus they provide a striking confirmation of the Einstein theory.

19. Black holes

The solution (18.6) becomes singular at $r = 2m$, because then $g_{00} = 0$ and $g_{11} = -\infty$. It would seem that $r = 2m$ gives a minimum radius for a body of mass m. But a closer investigation shows that this is not so.

Consider a particle falling into the central body and let its velocity vector be $v^\mu = dz^\mu/ds$. Let us suppose that it falls in radially, so that $v^2 = v^3 = 0$. The motion is determined by the geodesic equation (8.3):

$$\frac{dv^0}{ds} = -\Gamma^0_{\mu\nu} v^\mu v^\nu = -g^{00}\Gamma_{0\mu\nu} v^\mu v^\nu$$

$$= -g^{00} g_{00,1} v^0 v^1 = -g^{00} \frac{dg_{00}}{ds} v^0.$$

Now $g^{00} = 1/g_{00}$, so we get

$$g_{00} \frac{dv^0}{ds} + \frac{dg_{00}}{ds} v^0 = 0.$$

This integrates to

$$g_{00}v^0 = k,$$

with k a constant. It is the value of g_{00} where the particle starts to fall.

Again, we have

$$1 = g_{\mu\nu}v^\mu v^\nu = g_{00}v^{0^2} + g_{11}v^{1^2}.$$

Multiplying this equation by g_{00} and using $g_{00}g_{11} = -1$, which we obtained in the last section, we find

$$k^2 - v^{1^2} = g_{00} = 1 - \frac{2m}{r}.$$

For a falling body $v^1 < 0$, and hence

$$v^1 = -\left(k^2 - 1 + \frac{2m}{r}\right)^{1/2}.$$

Now

$$\frac{dt}{dr} = \frac{v^0}{v^1} = -k\left(1 - \frac{2m}{r}\right)^{-1}\left(k^2 - 1 + \frac{2m}{r}\right)^{-1/2}.$$

Let us suppose the particle is close to the critical radius, so $r = 2m + \varepsilon$ with ε small, and let us neglect ε^2. Then

$$\frac{dt}{dr} = -\frac{2m}{\varepsilon} = -\frac{2m}{r - 2m}.$$

This integrates to

$$t = -2m \log(r - 2m) + \text{constant}.$$

Thus, as $r \to 2m$, $t \to \infty$. The particle takes an infinite time to reach the critical radius $r = 2m$.

Let us suppose the particle is emitting light of a certain spectral line, and is being observed by someone at a large value of r. The light is red-shifted by a factor $g_{00}{}^{-1/2} = (1 - 2m/r)^{-1/2}$. This factor becomes infinite as the particle approaches the critical radius. All physical processes on the particle will be observed to be going more and more slowly as it approaches $r = 2m$.

Now consider an observer traveling with the particle. His time scale is measured by ds. Now

$$\frac{ds}{dr} = \frac{1}{v^1} = -\left(k^2 - 1 + \frac{2m}{r}\right)^{-1/2},$$

and this tends to $-k^{-1}$ as r tends to $2m$. Thus the particle reaches $r = 2m$ after the lapse of finite proper time for the observer. The traveling observer has aged only a finite amount when he reaches $r = 2m$. What will happen to him afterwards? He may continue sailing through empty space into smaller values of r.

To examine the continuation of the Schwarzschild solution for values of $r < 2m$, it is necessary to use a nonstatic system of coordinates, so that we have the $g_{\mu\nu}$ varying with the time coordinate. We keep the coordinates θ and ϕ unchanged, but instead of t and r we use τ and ρ, defined by

$$\tau = t + f(r), \qquad \rho = t + g(r), \tag{19.1}$$

where the functions f and g are at our disposal.

We have, using the prime again to denote the derivative with respect to r,

$$\begin{aligned} d\tau^2 - \frac{2m}{r} d\rho^2 &= (dt + f'\,dr)^2 - \frac{2m}{r}(dt + g'\,dr)^2 \\ &= \left(1 - \frac{2m}{r}\right) dt^2 + 2\left(f' - \frac{2m}{r} g'\right) dt\,dr + \left(f'^2 - \frac{2m}{r} g'^2\right) dr^2 \\ &= \left(1 - \frac{2m}{r}\right) dt^2 - \left(1 - \frac{2m}{r}\right)^{-1} dr^2, \end{aligned} \tag{19.2}$$

provided we choose the functions f and g to satisfy

$$f' = \frac{2m}{r} g' \tag{19.3}$$

and

$$\frac{2m}{r} g'^2 - f'^2 = \left(1 - \frac{2m}{r}\right)^{-1}. \tag{19.4}$$

Elimination of f from these equations gives

$$g' = \left(\frac{r}{2m}\right)^{1/2} \left(1 - \frac{2m}{r}\right)^{-1}. \tag{19.5}$$

To integrate this equation, put $r = y^2$ and $2m = a^2$. With $r > 2m$ we have $y > a$. We now have

$$\frac{dg}{dy} = 2y \frac{dg}{dr} = \frac{2y^4}{a} \frac{1}{y^2 - a^2},$$

which gives

$$g = \frac{2}{3a} y^3 + 2ay - a^2 \log \frac{y + a}{y - a}. \tag{19.6}$$

Finally, we get from (19.3) and (19.5)

$$g' - f' = \left(1 - \frac{2m}{r}\right)g' = \left(\frac{r}{2m}\right)^{1/2},$$

which integrates to

$$\frac{2}{3} \frac{1}{\sqrt{2m}} r^{3/2} = g - f = \rho - \tau. \tag{19.7}$$

Thus

$$r = \mu(\rho - \tau)^{2/3}, \tag{19.8}$$

with

$$\mu = (\tfrac{3}{2}\sqrt{2m})^{2/3}.$$

In this way we see that we can satisfy the conditions (19.3) and (19.4) and so we can use (19.2). Substituting into the Schwarzschild solution (18.6), we get

$$ds^2 = d\tau^2 - \frac{2m}{\mu(\rho - \tau)^{2/3}} d\rho^2 - \mu^2(\rho - \tau)^{4/3}(d\theta^2 + \sin^2 \theta \, d\phi^2). \tag{19.9}$$

The critical value $r = 2m$ corresponds, from (19.7), to $\rho - \tau = 4m/3$. There is no singularity here in the metric (19.9).

We know that the metric (19.9) satisfies the Einstein equations for empty space in the region $r > 2m$, because it can be transformed to the Schwarzschild solution by a mere change of coordinates. We can infer that it satisfies the Einstein equations also for $r \leq 2m$ from analytic continuity, because it does not involve any singularity at $r = 2m$. It may continue to hold right down to $r = 0$ or $\rho - \tau = 0$.

The singularity appears in the connection between the new coordinates and the original ones, equation (19.1). But once we have established the new coordinate system we can disregard the previous one and the singularity no longer appears.

We see that the Schwarzschild solution for empty space can be extended to the region $r < 2m$. But this region cannot communicate with the space for which $r > 2m$. Any signal, even a light signal, would take an infinite time to cross the boundary $r = 2m$, as we can easily check. Thus we cannot have direct observational knowledge of the region $r < 2m$. Such a region is called a black hole, because things may fall into it (taking an infinite time, by our clocks, to do so) but nothing can come out.

The question arises whether such a region can actually exist. All we can say definitely is that the Einstein equations allow it. A massive stellar object may collapse to a very small radius and the gravitational forces then become so strong that no known physical forces can hold them in check and prevent further collapse. It would seem that it would have to collapse into a black hole. It would take an infinite time to do so by our clocks, but only a finite time relatively to the collapsing matter itself.

20. Tensor densities

With a transformation of coordinates, an element of four-dimensional volume transforms according to the law

$$dx^{0'}\,dx^{1'}\,dx^{2'}\,dx^{3'} = dx^0\,dx^1\,dx^2\,dx^3 J, \tag{20.1}$$

where J is the Jacobian

$$J = \frac{\partial(x^{0'}x^{1'}x^{2'}x^{3'})}{\partial(x^0x^1x^2x^3)} = \text{determinant of } x^{\mu'}_{,\alpha}.$$

We may write (20.1)

$$d^4x' = J\,d^4x \tag{20.2}$$

for brevity.

Now

$$g_{\alpha\beta} = x^{\mu'}_{,\alpha}\,g_{\mu'\nu'}\,x^{\nu'}_{,\beta}.$$

We can look upon the right-hand side as the product of three matrices, the first matrix having its rows specified by α and columns specified by μ', the second having its rows specified by μ' and columns by ν', and the third having its rows specified by ν' and columns by β. This product equals the matrix $g_{\alpha\beta}$ on the left. The corresponding equation must hold between the determinants; therefore

$$g = Jg'J$$

or

$$g = J^2 g'.$$

Now g is a negative quantity, so we may form $\sqrt{-g}$, taking the positive value for the square root. Thus

$$\sqrt{-g} = J\sqrt{-g'}. \tag{20.3}$$

Suppose S is a scalar field quantity, $S = S'$. Then

$$\int S\sqrt{-g}\ d^4x = \int S\sqrt{-g'}\ J\ d^4x = \int S'\sqrt{-g'}\ d^4x',$$

if the region of integration for the x' corresponds to that for the x. Thus

$$\int S\sqrt{-g}\, d^4x = \text{invariant}. \tag{20.4}$$

We call $S\sqrt{-g}$ a scalar density, meaning a quantity whose integral is invariant.

Similarly, for any tensor field $T^{\mu\nu\cdots}$ we may call $T^{\mu\nu\cdots}\sqrt{-g}$ a tensor density. The integral

$$\int T^{\mu\nu}\sqrt{-g}\, d^4x$$

is a tensor if the domain of integration is small. It is not a tensor if the domain of integration is not small, because it then consists of a sum of tensors located at different points and it does not transform in any simple way under a transformation of coordinates.

The quantity $\sqrt{-g}$ will be very much used in the future. For brevity we shall write it simply as $\surd$. We have

$$g^{-1}g_{,\nu} = 2\surd^{-1}\surd_{,\nu}.$$

Thus the formula (14.5) gives

$$\surd_{,\nu} = \tfrac{1}{2}\surd g^{\lambda\mu}g_{\lambda\mu,\nu} \tag{20.5}$$

and the formula (14.6) may be written

$$\Gamma^{\mu}_{\nu\mu}\surd = \surd_{,\nu}. \tag{20.6}$$

21. Gauss and Stokes theorems

The vector A^μ has the covariant divergence $A^\mu{}_{:\mu}$, which is a scalar. We have

$$A^\mu{}_{:\mu} = A^\mu_{,\mu} + \Gamma^\mu_{\nu\mu} A^\nu = A^\mu_{,\mu} + \sqrt{}^{-1}\sqrt{}_{,\nu} A^\nu.$$

Thus

$$A^\mu{}_{:\mu}\sqrt{} = (A^\mu\sqrt{})_{,\mu}. \tag{21.1}$$

We can put $A^\mu{}_{:\mu}$ for S in (20.4), and we get the invariant

$$\int A^\mu{}_{:\mu}\sqrt{}\,d^4x = \int (A^\mu\sqrt{})_{,\mu}\, d^4x.$$

If the integral is taken over a finite (four-dimensional) volume, the right-hand side can be converted by Gauss's theorem to an integral over the boundary surface (three-dimensional) of the volume.

If $A^\mu{}_{:\mu} = 0$, we have

$$(A^\mu\sqrt{})_{,\mu} = 0 \tag{21.2}$$

and this gives us a conservation law; namely, the conservation of a fluid whose density is $A^0\sqrt{}$ and whose flow is given by the three-dimensional vector $A^m\sqrt{}$ ($m = 1, 2, 3$). We may integrate (21.2) over a three-dimensional volume V lying at a definite time x^0. The result is

$$\left(\int A^0\sqrt{}\,d^3x\right)_{,0} = -\int (A^m\sqrt{})_{,m}\, d^3x$$

$$= \text{surface integral over boundary of } V.$$

If there is no current crossing the boundary of V, $\int A^0\sqrt{}\, d^3x$ is constant.

These results for a vector A^μ cannot be taken over to a tensor with more than one suffix, in general. Take a two-suffix tensor $Y^{\mu\nu}$. In flat space we can use Gauss's theorem to express $\int Y^{\mu\nu}{}_{,\nu}\, d^4x$ as a surface integral, but in curved space we cannot in general express $\int Y^{\mu\nu}{}_{:\nu}\sqrt{}\, d^4x$ as a surface integral. An exception occurs for an antisymmetrical tensor $F^{\mu\nu} = -F^{\nu\mu}$.

In this case we have

$$F^{\mu\nu}{}_{:\sigma} = F^{\mu\nu}{}_{,\sigma} + \Gamma^\mu_{\sigma\rho}F^{\rho\nu} + \Gamma^\nu_{\sigma\rho}F^{\mu\rho},$$

so

$$F^{\mu\nu}{}_{:\nu} = F^{\mu\nu}{}_{,\nu} + \Gamma^\mu_{\nu\rho}F^{\rho\nu} + \Gamma^\nu_{\nu\rho}F^{\mu\rho}$$
$$= F^{\mu\nu}{}_{,\nu} + \sqrt{}^{-1}\sqrt{}_{,\rho}F^{\mu\rho}$$

from (20.6). Thus

$$F^{\mu\nu}{}_{:\nu}\sqrt{\ } = (F^{\mu\nu}\sqrt{\ })_{,\nu}. \tag{21.3}$$

Hence $\int F^{\mu\nu}{}_{:\nu}\sqrt{\ }d^4x$ = a surface integral, and if $F^{\mu\nu}{}_{:\nu} = 0$ we have a conservation law.

In the symmetrical case $Y^{\mu\nu} = Y^{\nu\mu}$ we can get a corresponding equation with an extra term, provided we put one of the suffixes downstairs and deal with $Y_\mu{}^\nu{}_{:\nu}$. We have

$$Y_\mu{}^\nu{}_{:\sigma} = Y_\mu{}^\nu{}_{,\sigma} - \Gamma^\alpha_{\mu\sigma}Y_\alpha{}^\nu + \Gamma^\nu_{\sigma\alpha}Y_\mu{}^\alpha.$$

Putting $\sigma = \nu$ and using (20.6), we get

$$Y_\mu{}^\nu{}_{:\nu} = Y_\mu{}^\nu{}_{,\nu} + \sqrt{\ }^{-1}\sqrt{\ }_{,\alpha}Y_\mu{}^\alpha - \Gamma_{\alpha\mu\nu}Y^{\alpha\nu}.$$

Since $Y^{\alpha\nu}$ is symmetrical, we can replace the $\Gamma_{\alpha\mu\nu}$ in the last term by

$$\tfrac{1}{2}(\Gamma_{\alpha\nu\mu} + \Gamma_{\nu\alpha\mu}) = \tfrac{1}{2}g_{\alpha\nu,\mu}$$

from (7.6). Thus we get

$$Y_\mu{}^\nu{}_{:\nu}\sqrt{\ } = (Y_\mu{}^\nu\sqrt{\ })_{,\nu} - \tfrac{1}{2}g_{\alpha\beta,\mu}Y^{\alpha\beta}\sqrt{\ }. \tag{21.4}$$

For a covariant vector A_μ, we have

$$\begin{aligned} A_{\mu:\nu} - A_{\nu:\mu} &= A_{\mu,\nu} - \Gamma^\rho_{\mu\nu}A_\rho - (A_{\nu,\mu} - \Gamma^\rho_{\nu\mu}A_\rho) \\ &= A_{\mu,\nu} - A_{\nu,\mu}. \end{aligned} \tag{21.5}$$

This result may be stated: covariant curl equals ordinary curl. It holds only for a covariant vector. For a contravariant vector we could not form the curl because the suffixes would not balance.

Let us take $\mu = 1$, $\nu = 2$. We get

$$A_{1:2} - A_{2:1} = A_{1,2} - A_{2,1}.$$

Let us integrate this equation over an area of the surface x^0 = constant, x^3 = constant. From Stokes's theorem we get

$$\begin{aligned} \iint (A_{1:2} - A_{2:1})\,dx^1\,dx^2 &= \iint (A_{1,2} - A_{2,1})\,dx^1\,dx^2 \\ &= \int (A_1\,dx^1 + A_2\,dx^2) \end{aligned} \tag{21.6}$$

integrated around the perimeter of the area. Thus we get an integral round a perimeter equated to a flux crossing the surface bounded by the perimeter.

The result must hold generally in all coordinate systems, not merely those for which the equations of the surface are x^0 = constant, x^3 = constant.

To get an invariant way of writing the result, we introduce a general formula for an element of two-dimensional surface. If we take two small contravariant vectors ξ^μ and ζ^μ, the element of surface area that they subtend is determined by the antisymmetric two-index tensor

$$dS^{\mu\nu} = \xi^\mu \zeta^\nu - \xi^\nu \zeta^\mu,$$

Thus, if ξ^μ has the components 0, dx^1, 0, 0, and ζ^μ has the components 0, 0, dx^2, 0, then $dS^{\mu\nu}$ has the components

$$dS^{12} = dx^1\, dx^2, \qquad dS^{21} = -dx^1\, dx^2,$$

with the other components vanishing. The left-hand side of (21.6) becomes

$$\iint A_{\mu:\nu}\, dS^{\mu\nu}.$$

The right-hand side is evidently $\int A_\mu\, dx^\mu$, so the formula becomes

$$\tfrac{1}{2} \iint_{\text{surface}} (A_{\mu:\nu} - A_{\nu:\mu})\, dS^{\mu\nu} = \int_{\text{perimeter}} A_\mu\, dx^\mu. \tag{21.7}$$

22. Harmonic coordinates

The d'Alembert equation for a scalar V, namely $\Box V = 0$, gives, from (10.9),

$$g^{\mu\nu}(V_{,\mu\nu} - \Gamma^\alpha_{\mu\nu} V_{,\alpha}) = 0. \tag{22.1}$$

If we are using rectilinear axes in flat space, each of the four coordinates x^λ satisfies $\Box x^\lambda = 0$. We might substitute x^λ for V in (22.1). The result, of course, is not a tensor equation, because x^λ is not a scalar like V, so it holds only in certain coordinate systems. It imposes a restriction on the coordinates.

If we substitute x^λ for V, then for $V_{,\alpha}$ we must substitute $x^\lambda_{,\alpha} = g^\lambda_\alpha$. The equation (22.1) becomes

$$g^{\mu\nu}\Gamma^\lambda_{\mu\nu} = 0. \tag{22.2}$$

Coordinates that satisfy this condition are called *harmonic coordinates.* They provide the closest approximation to rectilinear coordinates that we can have in curved space. We may use them in any problem if we wish to, but very often they are not worthwhile because the tensor formalism with general co-ordinates is really quite convenient. For the discussion of gravitational waves, however, harmonic coordinates are very useful.

We have in general coordinates, from (7.9) and (7.6),

$$\begin{aligned} g^{\mu\nu}{}_{,\sigma} &= -g^{\mu\alpha}g^{\nu\beta}(\Gamma_{\alpha\beta\sigma} + \Gamma_{\beta\alpha\sigma}) \\ &= -g^{\nu\beta}\Gamma^{\mu}_{\beta\sigma} - g^{\mu\alpha}\Gamma^{\nu}_{\alpha\sigma}. \end{aligned} \tag{22.3}$$

Thus, with the help of (20.6),

$$(g^{\mu\nu}\surd)_{,\sigma} = (-g^{\nu\beta}\Gamma^{\mu}_{\beta\sigma} - g^{\mu\alpha}\Gamma^{\nu}_{\alpha\sigma} + g^{\mu\nu}\Gamma^{\beta}_{\sigma\beta})\surd. \tag{22.4}$$

Contracting by putting $\sigma = \nu$, we get

$$(g^{\mu\nu}\surd)_{,\nu} = -g^{\nu\beta}\Gamma^{\mu}_{\beta\nu}\surd. \tag{22.5}$$

We see now that an alternative form for the harmonic condition is

$$(g^{\mu\nu}\surd)_{,\nu} = 0. \tag{22.6}$$

23. The electromagnetic field

Maxwell's equations, as ordinarily written, are

$$E = -\frac{1}{c}\frac{\partial A}{\partial t} - \operatorname{grad}\phi, \tag{23.1}$$

$$H = \operatorname{curl} A, \tag{23.2}$$

$$\frac{1}{c}\frac{\partial H}{\partial t} = -\operatorname{curl} E, \tag{23.3}$$

$$\operatorname{div} H = 0, \tag{23.4}$$

$$\frac{1}{c}\frac{\partial E}{\partial t} = \operatorname{curl} H - 4\pi j, \tag{23.5}$$

$$\operatorname{div} E = 4\pi\rho. \tag{23.6}$$

We must first put them in four-dimensional form for special relativity. The potentials A and ϕ form a four-vector κ^{μ} in accordance with

$$\kappa^0 = \phi, \qquad \kappa^m = A^m, \qquad (m = 1, 2, 3).$$

Define

$$F_{\mu\nu} = \kappa_{\mu,\nu} - \kappa_{\nu,\mu}. \tag{23.7}$$

Then from (23.1)

$$E^1 = -\frac{\partial \kappa^1}{\partial x^0} - \frac{\partial \kappa^0}{\partial x^1} = \frac{\partial \kappa_1}{\partial x^0} + \frac{\partial \kappa_0}{\partial x^1} = F_{10} = -F^{10}$$

and from (23.2)

$$H^1 = \frac{\partial \kappa^3}{\partial x^2} - \frac{\partial \kappa^2}{\partial x^3} = -\frac{\partial \kappa_3}{\partial x^2} + \frac{\partial \kappa_2}{\partial x^3} = F_{23} = F^{23}.$$

Thus the six components of the antisymmetric tensor $F_{\mu\nu}$ determine the field quantities E and H.

From the definition (23.7)

$$F_{\mu\nu,\sigma} + F_{\nu\sigma,\mu} + F_{\sigma\mu,\nu} = 0. \tag{23.8}$$

This gives the Maxwell equations (23.3) and (23.4). We have

$$F^{0\nu}{}_{,\nu} = F^{0m}{}_{m} = -F^{m0}{}_{m} = \operatorname{div} E = 4\pi\rho \tag{23.9}$$

from (23.6). Again

$$F^{1\nu}{}_{,\nu} = F^{10}{}_{,0} + F^{12}{}_{,2} + F^{13}{}_{,3} = -\frac{\partial E^1}{\partial x^0} + \frac{\partial H^3}{\partial x^2} - \frac{\partial H^2}{\partial x^3}$$
$$= 4\pi j^1. \tag{23.10}$$

from (23.5). The charge density ρ and current j^m form a four-vector J^{μ} in accordance with

$$J^0 = \rho, \qquad J^m = j^m.$$

Thus (23.9) and (23.10) combine into

$$F^{\mu\nu}{}_{,\nu} = 4\pi J^{\mu}. \tag{23.11}$$

In this way the Maxwell equations are put into the four-dimensional form required by special relativity.

To pass to general relativity we must write the equations in covariant form. On account of (21.5) we can write (23.7) immediately as

$$F_{\mu\nu} = \kappa_{\mu:\nu} - \kappa_{\nu:\mu}.$$

This gives us a covariant definition of the field quantities $F_{\mu\nu}$. We have further

$$F_{\mu\nu:\sigma} = F_{\mu\nu,\sigma} - \Gamma^{\alpha}_{\mu\sigma}F_{\alpha\nu} - \Gamma^{\alpha}_{\nu\sigma}F_{\mu\alpha}.$$

Making cyclic permutations of μ, ν, and σ and adding the three equations so obtained, we get

$$F_{\mu\nu:\sigma} + F_{\nu\sigma:\mu} + F_{\sigma\mu:\nu} = F_{\mu\nu,\sigma} + F_{\nu\sigma,\mu} + F_{\sigma\mu,\nu} = 0, \qquad (23.12)$$

from (23.8). So this Maxwell equation goes over immediately to the covariant form.

Finally, we must deal with the equation (23.11). This is not a valid equation in general relativity and must be replaced by the covariant equation

$$F^{\mu\nu}{}_{:\nu} = 4\pi J^{\mu}. \qquad (23.13)$$

From (21.3), which applies to any antisymmetric two-suffix tensor, we get

$$(F^{\mu\nu}\surd)_{,\nu} = 4\pi J^{\mu}\surd.$$

This leads immediately to

$$(J^{\mu}\surd)_{,\mu} = (4\pi)^{-1}(F^{\mu\nu}\surd)_{,\mu\nu} = 0.$$

So we have an equation like (21.2), giving us the law of conservation of electricity. The conservation of electricity holds accurately, undisturbed by the curvature of space.

24. Modification of the Einstein equations by the presence of matter

The Einstein equations in the absence of matter are

$$R^{\mu\nu} = 0. \qquad (24.1)$$

They lead to

$$R = 0;$$

and hence

$$R^{\mu\nu} - \tfrac{1}{2}g^{\mu\nu}R = 0. \tag{24.2}$$

If we start with equations (24.2), we get by contraction

$$R - 2R = 0$$

and so we can get back to (24.1). We may either use (24.1) or (24.2) as the basic equations for empty space.

In the presence of matter these equations must be modified. Let us suppose (24.1) is changed to

$$R^{\mu\nu} = X^{\mu\nu} \tag{24.3}$$

and (24.2) to

$$R^{\mu\nu} - \tfrac{1}{2}g^{\mu\nu}R = Y^{\mu\nu}. \tag{24.4}$$

Here $X^{\mu\nu}$ and $Y^{\mu\nu}$ are symmetric two-index tensors indicating the presence of matter.

We see now that (24.4) is the more convenient form to work with, because we have the Bianci relation (14.3), which tells us that

$$(R^{\mu\nu} - \tfrac{1}{2}g^{\mu\nu}R)_{:\nu} = 0.$$

Hence (24.4) requires

$$Y^{\mu\nu}{}_{:\nu} = 0. \tag{24.5}$$

Any tensor $Y^{\mu\nu}$ produced by matter must satisfy this condition; otherwise the equations (24.4) would not be consistent.

It is convenient to bring in the coefficient -8π and to rewrite equation (24.4) as

$$R^{\mu\nu} - \tfrac{1}{2}g^{\mu\nu}R = -8\pi Y^{\mu\nu}. \tag{24.6}$$

We shall find that the tensor $Y^{\mu\nu}$ with this coefficient is to be interpreted as the density and flux of (nongravitational) energy and momentum. $Y^{\mu 0}$ is the density and $Y^{\mu r}$ is the flux.

In flat space equation (24.5) would become

$$Y^{\mu\nu}{}_{,\nu} = 0$$

and would then give conservation of energy and momentum. In curved space the conservation of energy and momentum is only approximate. The error is to be ascribed to the gravitational field working on the matter and having itself some energy and momentum.

25. The material energy tensor

Suppose we have a distribution of matter whose velocity varies continuously from one point to a neighboring one. If z^μ denotes the coordinates of an element of the matter, we can introduce the velocity vector $v^\mu = dz^\mu/ds$, which will be a continuous function of the x's, like a field function. It has the properties

$$g_{\mu\nu} v^\mu v^\nu = 1, \tag{25.1}$$

$$0 = (g_{\mu\nu} v^\mu v^\nu)_{:\sigma} = g_{\mu\nu}(v^\mu v^\nu{}_{:\sigma} + v^\mu{}_{:\sigma} v^\nu)$$
$$= 2g_{\mu\nu} v^\mu v^\nu{}_{:\sigma}.$$

Thus

$$v_\nu v^\nu{}_{:\sigma} = 0. \tag{25.2}$$

We may introduce a scalar field ρ such that the vector field ρv^μ determines the density and flow of the matter just like J^μ determines the density and flow of electricity; that is to say, $\rho v^0 \surd$ is the density and $\rho v^m \surd$ is the flow. The condition for conservation of the matter is

$$(\rho v^\mu \surd)_{,\mu} = 0$$

or

$$(\rho v^\mu)_{:\mu} = 0. \tag{25.3}$$

The matter that we are considering will have an energy density $\rho v^0 v^0 \surd$ and energy flux $\rho v^0 v^m \surd$, and similarly a momentum density $\rho v^n v^0 \surd$ and momentum flux $\rho v^n v^m \surd$. Put

$$T^{\mu\nu} = \rho v^\mu v^\nu. \tag{25.4}$$

Then $T^{\mu\nu}\sqrt{}$ gives the density and flux of energy and momentum. $T^{\mu\nu}$ is called the material energy tensor. It is, of course, symmetric.

Can we use $T^{\mu\nu}$ for the matter term on the right-hand side of the Einstein equation (24.6)? For this purpose we require $T^{\mu\nu}{}_{:\nu} = 0$. We have from the definition (25.4)

$$T^{\mu\nu}{}_{:\nu} = (\rho v^\mu v^\nu)_{:\nu} = v^\mu(\rho v^\nu)_{:\nu} + \rho v^\nu v^\mu{}_{:\nu}.$$

The first term here vanishes from the condition for conservation of mass (25.3). The second term vanishes if the matter moves along geodesics for, if v^μ is defined as a continuous field function instead of having a meaning only on one world line, we have

$$\frac{dv^\mu}{ds} = v^\mu{}_{,\nu} v^\nu.$$

So (8.3) becomes

$$(v^\mu{}_{,\nu} + \Gamma^\mu_{\nu\sigma} v^\sigma) v^\nu = 0$$

or

$$v^\mu{}_{:\nu} v^\nu = 0. \tag{25.5}$$

We see now that we can substitute the material energy tensor (25.4), with a suitable numerical coefficient k, into the Einstein equation (24.4). We get

$$R^{\mu\nu} - \tfrac{1}{2} g^{\mu\nu} R = k\rho v^\mu v^\nu. \tag{25.6}$$

We shall now determine the value of the coefficient k. We go over to the Newtonian approximation, following the method of Section 16. We note first that, contracting (25.6), we get

$$-R = k\rho.$$

So (25.6) may be written

$$R^{\mu\nu} = k\rho(v^\mu v^\nu - \tfrac{1}{2} g^{\mu\nu}).$$

With the weak field approximation we get, corresponding to (16.4),

$$\tfrac{1}{2} g^{\rho\sigma}(g_{\rho\sigma,\mu\nu} - g_{\nu\sigma,\mu\rho} - g_{\mu\rho,\nu\sigma} + g_{\mu\nu,\rho\sigma}) = k\rho(v_\mu v_\nu - \tfrac{1}{2} g_{\mu\nu}).$$

We now take a static field and a static distribution of matter, so that $v_0 = 1$, $v_m = 0$. Putting $\mu = \nu = 0$ and neglecting second-order quantities, we find

$$-\tfrac{1}{2}\nabla^2 g_{00} = \tfrac{1}{2} k\rho$$

or from (16.6)

$$\nabla^2 V = -\tfrac{1}{2}k\rho.$$

To agree with the Poisson equation we must take $k = -8\pi$.

The Einstein equation for the presence of a distribution of matter with a velocity field thus reads

$$R^{\mu\nu} - \tfrac{1}{2}g^{\mu\nu}R = -8\pi\rho v^\mu v^\nu. \tag{25.7}$$

Thus $T^{\mu\nu}$, given by (25.4), is precisely the $Y^{\mu\nu}$ of equation (24.6).

The condition for conservation of mass (25.3) gives

$$\rho_{:\mu}v^\mu + \rho v^\mu{}_{:\mu} = 0;$$

hence

$$\frac{d\rho}{ds} = \frac{\partial\rho}{\partial x^\mu}v^\mu = -\rho v^\mu{}_{:\mu}. \tag{25.8}$$

This is a condition that fixes how ρ varies along the world line of an element of matter. It allows ρ to vary arbitrarily from the world line of one element to that of a neighboring element. Thus we may take ρ to vanish except for a packet of world lines forming a tube in space-time. Such a packet would compose a particle of matter of a finite size. Outside the particle we have $\rho = 0$, and Einstein's field equation for empty space holds.

It should be noted that, if one assumes the general field equation (25.7), one can deduce from it two things: (a) the mass is conserved and (b) the mass moves along geodesics. To do this we note that (left-hand side)$_{:\nu}$ vanishes from Bianci's relation, so the equation gives

$$(\rho v^\mu v^\nu)_{:\nu} = 0,$$

or

$$v^\mu(\rho v^\nu)_{:\nu} + \rho v^\nu v^\mu{}_{:\nu} = 0. \tag{25.9}$$

Multiply this equation by v_μ. The second term gives zero from (25.2) and we are left with $(\rho v^\nu)_{:\nu} = 0$, which is just the conservation equation (25.3). Equation (25.9) now reduces to $v^\nu v^\mu{}_{:\nu} = 0$, which is the geodesic equation. It is thus not necessary to make the separate assumption that a particle moves along a geodesic. With a small particle the motion is constrained to lie along a geodesic by the application of Einstein's equations for empty space to the space around the particle.

26. The gravitational action principle

Introduce the scalar

$$I = \int R\sqrt{\ }\, d^4x \tag{26.1}$$

integrated over a certain four-dimensional volume. Make small variations $\delta g_{\mu\nu}$ in the $g_{\mu\nu}$, keeping the $g_{\mu\nu}$ and their first derivatives constant on the boundary. We shall find that putting $\delta I = 0$ for arbitrary $\delta g_{\mu\nu}$ gives Einstein's vacuum equations.

We have from (14.4)

$$R = g^{\mu\nu}R_{\mu\nu} = R^* - L,$$

where

$$R^* = g^{\mu\nu}(\Gamma^{\sigma}_{\mu\sigma,\nu} - \Gamma^{\sigma}_{\mu\nu,\sigma}) \tag{26.2}$$

and

$$L = g^{\mu\nu}(\Gamma^{\sigma}_{\mu\nu}\Gamma^{\rho}_{\sigma\rho} - \Gamma^{\rho}_{\mu\sigma}\Gamma^{\sigma}_{\nu\rho}). \tag{26.3}$$

I involves second derivatives of $g_{\mu\nu}$, since these second derivatives occur in R^*. But they occur only linearly, so they can be removed by partial integration. We have

$$R^*\sqrt{\ } = (g^{\mu\nu}\Gamma^{\sigma}_{\mu\sigma}\sqrt{\ })_{,\nu} - (g^{\mu\nu}\Gamma^{\sigma}_{\mu\nu}\sqrt{\ })_{,\sigma} - (g^{\mu\nu}\sqrt{\ })_{,\nu}\Gamma^{\sigma}_{\mu\sigma} + (g^{\mu\nu}\sqrt{\ })_{,\sigma}\Gamma^{\sigma}_{\mu\nu}. \tag{26.4}$$

The first two terms are perfect differentials, so they will contribute nothing to I. We therefore need retain only the last two terms of (26.4). With the help of (22.5) and (22.4) they become

$$g^{\nu\beta}\Gamma^{\mu}_{\beta\nu}\Gamma^{\sigma}_{\mu\sigma}\sqrt{\ } + (-2g^{\nu\beta}\Gamma^{\mu}_{\beta\sigma} + g^{\mu\nu}\Gamma^{\beta}_{\sigma\beta})\Gamma^{\sigma}_{\mu\nu}\sqrt{\ }.$$

This is just $2L\sqrt{\ }$, from (26.3). So (26.1) becomes

$$I = \int L\sqrt{\ }\, d^4x,$$

which involves only the $g_{\mu\nu}$ and their first derivatives. It is homogeneous of the second degree in these first derivatives.

Put $\mathscr{L} = L\sqrt{\ }$. We take it (with a suitable numerical coefficient to be determined later) as the action density for the gravitational field. It is not a scalar density. But it is more convenient than $R\surd$, which is a scalar density, because it does not involve second derivatives of the $g_{\mu\nu}$.

According to ordinary ideas of dynamics, the action is the time integral of the Lagrangian. We have

$$I = \int \mathscr{L}\, d^4x = \int dx_0 \int \mathscr{L}\, dx^1\, dx^2\, dx^3$$

so the Lagrangian is evidently

$$\int \mathscr{L}\, dx^1\, dx^2\, dx^3.$$

Thus $\mathscr{L}$ may be considered as the Lagrangian density (in three dimensions) as well as the action density (in four dimensions). We may look upon the $g_{\mu\nu}$ as dynamical coordinates and their time derivatives as the velocities. We then see that the Lagrangian is quadratic (nonhomogeneous) in the velocities, as it usually is in ordinary dynamics.

We must now vary $\mathscr{L}$. We have, using (20.6),

$$\begin{aligned}\delta(\Gamma^{\alpha}_{\mu\nu}\Gamma^{\beta}_{\alpha\beta}g^{\mu\nu}\surd) &= \Gamma^{\alpha}_{\mu\nu}\,\delta(\Gamma^{\beta}_{\alpha\beta}g^{\mu\nu}\surd) + \Gamma^{\beta}_{\alpha\beta}g^{\mu\nu}\surd\,\delta\Gamma^{\alpha}_{\mu\nu}\\ &= \Gamma^{\alpha}_{\mu\nu}\,\delta(g^{\mu\nu}\surd_{,\alpha}) + \Gamma^{\beta}_{\alpha\beta}\,\delta(\Gamma^{\alpha}_{\mu\nu}g^{\mu\nu}\surd) - \Gamma^{\beta}_{\alpha\beta}\Gamma^{\alpha}_{\mu\nu}\,\delta(g^{\mu\nu}\surd)\\ &= \Gamma^{\alpha}_{\mu\nu}\,\delta(g^{\mu\nu}\surd_{,\alpha}) - \Gamma^{\beta}_{\alpha\beta}\,\delta(g^{\alpha\nu}\surd)_{,\nu} - \Gamma^{\beta}_{\alpha\beta}\Gamma^{\alpha}_{\mu\nu}\,\delta(g^{\mu\nu}\surd)\end{aligned} \tag{26.5}$$

with the help of (22.5). Again

$$\begin{aligned}\delta(\Gamma^{\beta}_{\mu\alpha}\Gamma^{\alpha}_{\nu\beta}g^{\mu\nu}\surd) &= 2(\delta\Gamma^{\beta}_{\mu\alpha})\Gamma^{\alpha}_{\nu\beta}g^{\mu\nu}\surd + \Gamma^{\beta}_{\mu\alpha}\Gamma^{\alpha}_{\nu\beta}\,\delta(g^{\mu\nu}\surd)\\ &= 2\delta(\Gamma^{\beta}_{\mu\alpha}g^{\mu\nu}\surd)\Gamma^{\alpha}_{\nu\beta} - \Gamma^{\beta}_{\mu\alpha}\Gamma^{\alpha}_{\nu\beta}\,\delta(g^{\mu\nu}\surd)\\ &= -\delta(g^{\nu\beta}{}_{,\alpha}\surd)\Gamma^{\alpha}_{\nu\beta} - \Gamma^{\beta}_{\mu\alpha}\Gamma^{\alpha}_{\nu\beta}\delta(g^{\mu\nu}\surd)\end{aligned} \tag{26.6}$$

with the help of (22.3). Subtracting (26.6) from (26.5), we get

$$\delta\mathscr{L} = \Gamma^{\alpha}_{\mu\nu}\,\delta(g^{\mu\nu}\surd)_{,\alpha} - \Gamma^{\beta}_{\alpha\beta}\,\delta(g^{\alpha\nu}\surd)_{,\nu} + (\Gamma^{\beta}_{\mu\alpha}\Gamma^{\alpha}_{\nu\beta} - \Gamma^{\beta}_{\alpha\beta}\Gamma^{\alpha}_{\mu\nu})\,\delta(g^{\mu\nu}\surd). \tag{26.7}$$

The first two terms here differ by a perfect differential from

$$-\Gamma^{\alpha}_{\mu\nu,\alpha}\,\delta(g^{\mu\nu}\surd) + \Gamma^{\beta}_{\mu\beta,\nu}\,\delta(g^{\mu\nu}\surd).$$

So we get

$$\delta I = \delta\int \mathscr{L}\, d^4x = \int R_{\mu\nu}\,\delta(g^{\mu\nu}\surd)\, d^4x, \tag{26.8}$$

with $R_{\mu\nu}$ given by (14.4). With the $\delta g_{\mu\nu}$ arbitrary, the quantities $\delta(g^{\mu\nu}\surd)$ are also independent and arbitrary, so the condition that (26.8) vanishes leads to Einstein's law in the form (24.1).

We can deduce, by the same method as (7.9), that

$$\delta g^{\mu\nu} = -g^{\mu\alpha}g^{\nu\beta}\,\delta g_{\alpha\beta}. \tag{26.9}$$

Also, corresponding to (20.5), we can deduce

$$\delta\surd = \tfrac{1}{2}\surd g^{\alpha\beta}\,\delta g_{\alpha\beta}. \tag{26.10}$$

Thus

$$\delta(g^{\mu\nu}\surd) = -(g^{\mu\alpha}g^{\nu\beta} - \tfrac{1}{2}g^{\mu\nu}g^{\alpha\beta})\surd\,\delta g_{\alpha\beta}.$$

So we may write (26.8), alternatively,

$$\delta I = -\int R_{\mu\nu}(g^{\mu\alpha}g^{\nu\beta} - \tfrac{1}{2}g^{\mu\nu}g^{\alpha\beta})\surd\,\delta g_{\alpha\beta}\,d^4x$$

$$= -\int (R^{\alpha\beta} - \tfrac{1}{2}g^{\alpha\beta}R)\surd\,\delta g_{\alpha\beta}\,d^4x. \tag{26.11}$$

The requirement that (26.11) vanishes gives Einstein's law in the form (24.2).

27. The action for a continuous distribution of matter

We shall consider a continuous distribution of matter whose velocity varies continuously from one point to a neighboring one, as we did in Section 25. We shall set up an action principle for this matter in interaction with the gravitational field in the form

$$\delta(I_g + I_m) = 0, \tag{27.1}$$

where I_g, the gravitational part of the action, is the I of the preceding section with some numerical coefficient κ, and I_m, the matter part of the action, will now be determined. The condition (27.1) must lead to Einstein's equations (25.7) for the gravitational field in the presence of the matter and the geodesic equations of motion for the matter.

We shall need to make arbitrary variations in the position of an element of matter to see how it affects I_m. It makes the discussion clearer if we first consider the variations purely kinematically, without any reference to the metric $g_{\mu\nu}$. There is then a real distinction between covariant and contravariant vectors and we cannot transform one into the other. A velocity is described by the ratios of the components of a contravariant vector u^μ, and it cannot be normalized without bringing in the metric.

With a continuous flow of matter we have a velocity vector u^μ (with an unknown multiplying factor) at each point. We can set up a contravariant vector density p^μ, lying in the direction of u^μ, which determines both the quantity of the flow and its velocity according to the formulas:

$$p^0\, dx^1\, dx^2\, dx^3$$

is the amount of matter within the element of volume $dx^1\, dx^2\, dx^3$ at a certain time and

$$p^1\, dx^0\, dx^2\, dx^3$$

is the amount flowing through the surface element $dx^2\, dx^3$ during a time interval dx^0. We shall assume the matter is conserved, so

$$p^\mu{}_{,\mu} = 0. \tag{27.2}$$

Let us suppose each element of matter is displaced from z^μ to $z^\mu + b^\mu$ with b^μ small. We must determine the resulting change in p^μ at a given point x.

Take first the case of $b^0 = 0$. The change in the amount of matter within a certain three-dimensional volume V is minus the amount displaced through the boundary of V:

$$\delta \int_V p^0\, dx^1\, dx^2\, dx^3 = -\int p^0 b^r\, dS_r,$$

($r = 1, 2, 3$), where dS_r denotes an element of the boundary surface of V. We can transform the right-hand side to a volume integral by Gauss's theorem and we find

$$\delta p^0 = -(p^0 b^r)_{,r}. \tag{27.3}$$

We must generalize this result to the case $b^0 \neq 0$. We make use of the condition that if b^μ is proportional to p^μ, each element of matter is displaced along its world line and there is then no change in p^μ. The generalization of (27.3) is evidently

$$\delta p^0 = (p^r b^0 - p^0 b^r)_{,r}$$

because this agrees with (27.3) when $b^0 = 0$ and gives $\delta p^0 = 0$ when b^μ is proportional to p^μ. There is a corresponding formula for the other components of p^μ, so the general result is

$$\delta p^\mu = (p^\nu b^\mu - p^\mu b^\nu)_{,\nu}. \tag{27.4}$$

For describing a continuous flow of matter the quantities p^μ are the basic variables to be used in the action function. They must be varied in accordance with the formula (27.4), and then, after suitable partial integrations, we must put the coefficient of each b^μ equal to zero. This will give us the equations of motion for the matter.

The action for an isolated particle of mass m is

$$-m \int ds. \tag{27.5}$$

We see the need for the coefficient $-m$ by taking the case of special relativity, for which the Lagrangian would be the time derivative of (27.5), namely

$$L = -m \frac{ds}{dx^0} = -m \left(1 - \frac{dx^r}{dx^0}\frac{dx^r}{dx^0}\right)^{1/2},$$

summed for $r = 1, 2, 3$. This gives for the momentum

$$\frac{\partial L}{\partial(dx^r/dx^0)} = m \frac{dx^r}{dx^0}\left(1 - \frac{dx^n}{dx^0}\frac{dx^n}{dx^0}\right)^{-1/2}$$

$$= m \frac{dx^r}{ds},$$

as it ought to be.

We obtain the action for a continuous distribution of matter from (27.5) by replacing m by $p^0\, dx^1\, dx^2\, dx^3$ and integrating; thus

$$I_m = -\int p^0\, dx^1\, dx^2\, dx^3\, ds. \tag{27.6}$$

To get this in a more understandable form we use the metric and put

$$p^\mu = \rho v^\mu \surd, \tag{27.7}$$

where ρ is a scalar that determines the density and v^μ is the previous vector u^μ normalized to be of length 1. We get

$$I_m = -\int \rho \sqrt{\ } v^0 \, dx^1 \, dx^2 \, dx^3 \, ds$$
$$= -\int \rho \sqrt{\ } \, d^4x, \tag{27.8}$$

since $v^0 \, ds = dx^0$.

This form for the action is not suitable for applying variations, because ρ, v^μ are not independent variables. We must eliminate them in terms of the p^μ, which are then to be varied in accordance with (27.4). We get from (27.7)

$$(p^\mu p_\mu)^{1/2} = \rho \sqrt{\ }.$$

So (27.8) becomes

$$I_m = -\int (p^\mu p_\mu)^{1/2} \, d^4x. \tag{27.9}$$

To vary this expression we use

$$\delta(p^\mu p_\mu)^{1/2} = \tfrac{1}{2}(p^\lambda p_\lambda)^{-1/2}(p^\mu p^\nu \, \delta g_{\mu\nu} + 2p_\mu \, \delta p^\mu)$$
$$= \tfrac{1}{2}\rho v^\mu v^\nu \sqrt{\ } \, \delta g_{\mu\nu} + v_\mu \, \delta p^\mu.$$

The action principle (27.1) now gives, with the help of (26.11), which we multiply by the coefficient κ,

$$\delta(I_g + I_m) = -\int [\kappa(R^{\mu\nu} - \tfrac{1}{2}g^{\mu\nu}R) + \tfrac{1}{2}\rho v^\mu v^\nu] \sqrt{\ } \, \delta g_{\mu\nu} \, d^4x - \int v_\mu \, \delta p^\mu \, d^4x. \tag{27.10}$$

Equating to zero the coefficient of $\delta g_{\mu\nu}$, we get Einstein's equation (25.7), provided we take $\kappa = (16\pi)^{-1}$. The last term gives, with (27.4)

$$-\int v_\mu (p^\nu b^\mu - p^\mu b^\nu)_{,\nu} \, d^4x$$
$$= \int v_{\mu,\nu}(p^\nu b^\mu - p^\mu b^\nu) \, d^4x$$
$$= \int (v_{\mu,\nu} - v_{\nu,\mu}) p^\nu b^\mu \, d^4x$$
$$= \int (v_{\mu:\nu} - v_{\nu:\mu}) \rho v^\nu b^\mu \sqrt{\ } \, d^4x$$
$$= \int v_{\mu:\nu} \rho v^\nu b^\mu \sqrt{\ } \, d^4x \tag{27.11}$$

from (25.2). Equating to zero the coefficient of b^μ here, we get the geodesic equation (25.5).

28. The action for the electromagnetic field

The usual expression for the action density of the electromagnetic field is

$$(8\pi)^{-1}(E^2 - H^2).$$

If we write it in the four-dimensional notation of special relativity given in Section 23, it becomes

$$-(16\pi)^{-1}F_{\mu\nu}F^{\mu\nu}.$$

This leads to the expression

$$I_{em} = -(16\pi)^{-1}\int F_{\mu\nu}F^{\mu\nu}\surd\, d^4x \tag{28.1}$$

for the invariant action in general relativity. Here we must take into account that $F_{\mu\nu} = \kappa_{\mu,\nu} - \kappa_{\nu,\mu}$, so I_{em} is a function of the $g_{\mu\nu}$ and the derivatives of the electromagnetic potentials.

Let us first vary the $g_{\mu\nu}$, keeping the κ_σ constant, so the $F_{\mu\nu}$ are constant but not the $F^{\mu\nu}$. We have

$$\begin{aligned}\delta(F_{\mu\nu}F^{\mu\nu}\surd) &= F_{\mu\nu}F^{\mu\nu}\,\delta\surd + F_{\mu\nu}F_{\alpha\beta}\surd\,\delta(g^{\mu\alpha}g^{\nu\beta}) \\ &= \tfrac{1}{2}F_{\mu\nu}F^{\mu\nu}g^{\rho\sigma}\surd\,\delta g_{\rho\sigma} - 2F_{\mu\nu}F_{\alpha\beta}\surd g^{\mu\rho}g^{\alpha\sigma}g^{\nu\beta}\,\delta g_{\rho\sigma}\end{aligned}$$

with the help of (26.10) and (26.9). Thus

$$\begin{aligned}\delta(F_{\mu\nu}F^{\mu\nu}\surd) &= (\tfrac{1}{2}F_{\mu\nu}F^{\mu\nu}g^{\rho\sigma} - 2F^{\rho}{}_{\nu}F^{\sigma\nu})\surd\,\delta g_{\rho\sigma} \\ &= 8\pi E^{\rho\sigma}\surd\,\delta g_{\rho\sigma},\end{aligned} \tag{28.2}$$

where $E^{\rho\sigma}$ is the stress-energy tensor of the electromagnetic field, a symmetrical tensor defined by

$$4\pi E^{\rho\sigma} = -F^{\rho}{}_{\nu}F^{\sigma\nu} + \tfrac{1}{4}g^{\rho\sigma}F_{\mu\nu}F^{\mu\nu}. \tag{28.3}$$

Note that in special relativity

$$\begin{aligned} 4\pi E^{00} &= E^2 - \tfrac{1}{2}(E^2 - H^2) \\ &= \tfrac{1}{2}(E^2 + H^2), \end{aligned}$$

so E^{00} is the energy density, and

$$\begin{aligned} 4\pi E^{01} &= -F^0{}_2 F^{12} - F^0{}_3 F^{13} \\ &= E^2 H^3 - E^3 H^2, \end{aligned}$$

so E^{0n} is the Poynting vector giving the rate of flow of energy.

If we vary the κ_μ, keeping the $g_{\alpha\beta}$ fixed, we get

$$\begin{aligned} \delta(F_{\mu\nu} F^{\mu\nu} \surd) &= 2F^{\mu\nu} \surd\, \delta F_{\mu\nu} = 4F^{\mu\nu} \surd\, \delta\kappa_{\mu,\nu} \\ &= 4(F^{\mu\nu} \surd\, \delta\kappa_\mu)_{,\nu} - 4(F^{\mu\nu} \surd)_{,\nu}\, \delta\kappa_\mu \\ &= 4(F^{\mu\nu} \surd\, \delta\kappa_\mu)_{,\nu} - 4F^{\mu\nu}{}_{:\nu} \surd\, \delta\kappa_\mu \end{aligned} \tag{28.4}$$

with the help of (21.3).

Adding (28.2) and (28.4) and dividing by -16π, we get for the total variation

$$\delta I_{em} = \int [-\tfrac{1}{2} E^{\mu\nu}\, \delta g_{\mu\nu} + (4\pi)^{-1} F^{\mu\nu}{}_{:\nu}\, \delta\kappa_\mu] \surd\, d^4x. \tag{28.5}$$

29. The action for charged matter

In the preceding section we considered the electromagnetic field in the absence of charges. If there are charges present, a further term is needed in the action. For a single particle of charge e, the extra action is

$$-e \int \kappa_\mu\, dx^\mu = -e \int \kappa_\mu v^\mu\, ds, \tag{29.1}$$

integrated along the world line.

There are difficulties in dealing with a point particle carrying a charge because it produces a singularity in the electric field. We can evade these difficulties by dealing instead with a continuous distribution of matter carrying charge. We shall handle this matter with the technique of Section 27, assuming each element of the matter carries charge.

In the kinematical discussion we had a contravariant vector density p^μ to determine the density and flow of the matter. We must now introduce a contravariant vector density $\mathcal{J}^\mu$ to determine the density and flow of electricity. The two vectors are constrained to lie in the same direction. When we make a displacement, we have

$$\delta \mathcal{J}^\mu = (\mathcal{J}^\nu b^\mu - \mathcal{J}^\mu b^\nu)_{,\nu} \tag{29.2}$$

corresponding to (27.4), with the same b^μ.

The expression (29.1) for the action for a charged particle now leads to

$$I_q = -\int \mathcal{J}^0 \kappa_\mu v^\mu \, dx^1 \, dx^2 \, dx^3 \, ds$$

for a continuous distribution of charged matter, corresponding to (27.6).

When we introduce the metric we put, corresponding to (27.7),

$$\mathcal{J}^\mu = \sigma v^\mu \sqrt{\ }, \tag{29.3}$$

where σ is a scalar that determines the charge density. The action now becomes, corresponding to (27.8),

$$\begin{aligned} I_q &= -\int \sigma \kappa_\mu v^\mu \sqrt{\ } \, d^4x \\ &= -\int \kappa_\mu \mathcal{J}^\mu \, d^4x. \end{aligned} \tag{29.4}$$

Thus

$$\begin{aligned} \delta I_q &= -\int [\mathcal{J}^\mu \, \delta\kappa_\mu + \kappa_\mu (\mathcal{J}^\nu b^\mu - \mathcal{J}^\mu b^\nu)_{,\nu}] \, d^4x \\ &= \int [-\sigma v^\mu \sqrt{\ } \, \delta\kappa_\mu + \kappa_{\mu,\nu} (\mathcal{J}^\nu b^\mu - \mathcal{J}^\mu b^\nu)] \, d^4x \\ &= \int \sigma(-v^\mu \, \delta\kappa_\mu + F_{\mu\nu} v^\nu b^\mu) \sqrt{\ } \, d^4x. \end{aligned} \tag{29.5}$$

The equations for the interaction of the charged matter with the combined gravitational and electromagnetic fields all follow from the general action principle

$$\delta(I_g + I_m + I_{em} + I_q) = 0. \tag{29.6}$$

Thus we take the sum of the expressions (29.5), (28.5), and (27.10) with the last term replaced by (27.11), and equate the total coefficients of the variations $\delta g_{\mu\nu}$, $\delta\kappa_\mu$, and b^μ to zero.

The coefficient of $\sqrt{\ }\,\delta g_{\mu\nu}$, multiplied by -16π, gives

$$R^{\mu\nu} - \tfrac{1}{2}g^{\mu\nu}R + 8\pi\rho v^\mu v^\nu + 8\pi E^{\mu\nu} = 0. \tag{29.7}$$

This is the Einstein equation (24.6) with $Y^{\mu\nu}$ consisting of two parts, one coming from the material-energy tensor and the other from the stress-energy tensor of the electromagnetic field.

The coefficient of $\sqrt{\ }\,\delta\kappa_\mu$ gives

$$-\sigma v^\mu + (4\pi)^{-1}F^{\mu\nu}{}_{:\nu} = 0.$$

From (29.3) we see that σv^μ is the charge current vector J^μ, so we get

$$F^{\mu\nu}{}_{:\nu} = 4\pi J^\mu. \tag{29.8}$$

This is the Maxwell equation (23.13) for the presence of charges.

Finally, the coefficient of $\sqrt{\ }b^\mu$ gives

$$\rho v_{\mu:\nu}v^\nu + \sigma F_{\mu\nu}v^\nu = 0,$$

or

$$\rho v_{\mu:\nu}v^\nu + F_{\mu\nu}J^\nu = 0. \tag{29.9}$$

The second term here gives the Lorentz force which causes the trajectory of an element of the matter to depart from a geodesic.

The equation (29.9) can be deduced from (29.7) and (29.8). Taking the covariant divergence of (29.7) and using the Bianci relations, we get

$$(\rho v^\mu v^\nu + E^{\mu\nu})_{:\nu} = 0. \tag{29.10}$$

Now from (28.3)

$$\begin{aligned} 4\pi E^{\mu\nu}{}_{:\nu} &= -F^{\mu\alpha}F^{\nu}{}_{\alpha:\nu} - F^{\mu\alpha}{}_{:\nu}F^{\nu}{}_{\alpha} + \tfrac{1}{2}g^{\mu\nu}F^{\alpha\beta}F_{\alpha\beta:\nu} \\ &= -F^{\mu\alpha}F^{\nu}{}_{\alpha:\nu} - \tfrac{1}{2}g^{\mu\rho}F^{\nu\sigma}(F_{\rho\sigma:\nu} - F_{\rho\nu:\sigma} - F_{\nu\sigma:\rho}) \\ &= 4\pi F^{\mu\alpha}J_\alpha, \end{aligned}$$

from (23.12) and (29.8). So (29.10) becomes

$$v^\mu(\rho v^\nu)_{:\nu} + \rho v^\nu v^\mu{}_{:\nu} + F^{\mu\alpha}J_\alpha = 0. \tag{29.11}$$

Multiply by v_μ and use (25.2). We get

$$(\rho v^\nu)_{:\nu} = -F^{\mu\alpha}v_\mu J_\alpha = 0$$

if we use the condition $J_\alpha = \sigma v_\alpha$, expressing that J_α and v_α are constrained to lie in the same direction. Thus the first term of (29.11) vanishes and we are left with (29.9).

This deduction means that the equations that follow from the action principle (29.6) are not all independent. There is a general reason for this, which will be explained in Section 30.

30. The comprehensive action principle

The method of Section 29 can be generalized to apply to the gravitational field interacting with any other fields, which are also interacting with one another. There is a comprehensive action principle,

$$\delta(I_g + I') = 0, \tag{30.1}$$

where I_g is the gravitational action that we had before and I' is the action of all the other fields and consists of a sum of terms, one for each field. It is a great advantage of using an action principle that it is so easy to obtain the correct equations for any fields in interaction. One merely has to obtain the action for each of the fields one is interested in and add them all together and include them all in (30.1).

We have

$$I_g = \int \mathscr{L}\, d^4x,$$

where this $\mathscr{L}$ is $(16\pi)^{-1}$ times the $\mathscr{L}$ of Section 26. We get

$$\begin{aligned} \delta I_g &= \int \left(\frac{\partial \mathscr{L}}{\partial g_{\alpha\beta}} \delta g_{\alpha\beta} + \frac{\partial \mathscr{L}}{\partial g_{\alpha\beta,\nu}} \delta g_{\alpha\beta,\nu} \right) d^4x \\ &= \int \left[\frac{\partial \mathscr{L}}{\partial g_{\alpha\beta}} - \left(\frac{\partial \mathscr{L}}{\partial g_{\alpha\beta,\nu}} \right)_{,\nu} \right] \delta g_{\alpha\beta}\, d^4x. \end{aligned}$$

The work of Section 26, leading to (26.11), shows that

$$\frac{\partial \mathscr{L}}{\partial g_{\alpha\beta}} - \left(\frac{\partial \mathscr{L}}{\partial g_{\alpha\beta,\nu}} \right)_{,\nu} = -(16\pi)^{-1}(R^{\alpha\beta} - \tfrac{1}{2}g^{\alpha\beta}R)\surd. \tag{30.2}$$

Let ϕ_n $(n = 1, 2, 3, \ldots)$ denote the other field quantities. Each of them is assumed to be a component of a tensor, but its precise tensor character is left unspecified. I' is of the form of the integral of a scalar density

$$I' = \int \mathscr{L}' \, d^4x,$$

where $\mathscr{L}'$ is a function of the ϕ_n and their first derivatives $\phi_{n,\mu}$ and possibly also higher derivatives.

The variation of the action now leads to a result of the form

$$\delta(I_g + I') = \int (p^{\mu\nu} \, \delta g_{\mu\nu} + \Sigma_n \chi^n \, \delta\phi_n)\sqrt{\ } \, d^4x, \tag{30.3}$$

with $p^{\mu\nu} = p^{\nu\mu}$, because any term involving δ (derivative of a field quantity) can be transformed by partial integration to a term that can be included in (30.3). The variation principle (30.1) thus leads to the field equations

$$p^{\mu\nu} = 0, \tag{30.4}$$

$$\chi^n = 0. \tag{30.5}$$

$p^{\mu\nu}$ will consist of the term (30.2) coming from I_g plus terms coming from $\mathscr{L}'$, say $N^{\mu\nu}$. We have of course $N^{\mu\nu} = N^{\nu\mu}$. $\mathscr{L}'$ usually does not contain derivatives of the $g_{\mu\nu}$ and then

$$N^{\mu\nu} = \frac{\partial \mathscr{L}'}{\partial g_{\mu\nu}}. \tag{30.6}$$

The equation (30.4) now becomes

$$R^{\mu\nu} - \tfrac{1}{2}g^{\mu\nu}R - 16\pi N^{\mu\nu} = 0.$$

It is just the Einstein equation (24.6) with

$$Y^{\mu\nu} = -2N^{\mu\nu}. \tag{30.7}$$

We see here how each field contributes a term to the right-hand side of the Einstein equation, depending, according to (30.6), on the way the action for that field involves $g_{\mu\nu}$.

It is necessary for consistency that the $N^{\mu\nu}$ have the property $N^{\mu\nu}{}_{:\nu} = 0$. This property can be deduced quite generally from the condition that I' is invariant under a change of coordinates that leaves the bounding surface unchanged. We make a small change of coordinates, say $x^{\mu'} = x^\mu + b^\mu$, with the b^μ small and functions of the x's, and work to the first order in the b^μ. The transformation law for the $g_{\mu\nu}$ is according to (3.7), with dashed suffixes to specify the new tensor,

$$g_{\mu\nu}(x) = x^{\alpha'}_{,\mu} x^{\beta'}_{,\nu} g_{\alpha'\beta'}(x'). \tag{30.8}$$

Let $\delta g_{\alpha\beta}$ denote the first-order change in $g_{\alpha\beta}$, not at a specified field point, but for definite values of the coordinates to which it refers, so that

$$\begin{aligned} g_{\alpha'\beta'}(x') &= g_{\alpha\beta}(x') + \delta g_{\alpha\beta} \\ &= g_{\alpha\beta}(x) + g_{\alpha\beta,\sigma} b^\sigma + \delta g_{\alpha\beta}. \end{aligned}$$

We have

$$x^{\alpha'}_{,\mu} = (x^\alpha + b^\alpha)_{,\mu} = g^\alpha_\mu + b^\alpha_{,\mu}.$$

Thus (30.8) gives

$$\begin{aligned} g_{\mu\nu}(x) &= (g^\alpha_\mu + b^\alpha_{,\mu})(g^\beta_\nu + b^\beta_{,\nu})[g_{\alpha\beta}(x) + g_{\alpha\beta,\sigma} b^\sigma + \delta g_{\alpha\beta}] \\ &= g_{\mu\nu}(x) + g_{\mu\nu,\sigma} b^\sigma + \delta g_{\mu\nu} + g_{\mu\beta} b^\beta_{,\nu} + g_{\alpha\nu} b^\alpha_{,\mu}, \end{aligned}$$

so

$$\delta g_{\mu\nu} = -g_{\mu\alpha} b^\alpha_{,\nu} - g_{\nu\alpha} b^\alpha_{,\mu} - g_{\mu\nu,\sigma} b^\sigma.$$

We now determine the variation in I' when the $g_{\mu\nu}$ are changed in this way and the other field variables keep the same value at the point with coordinates $x^{\mu'}$ that they previously had for x^μ. It is, if we use (30.6),

$$\begin{aligned} \delta I' &= \int N^{\mu\nu} \delta g_{\mu\nu} \sqrt{\ } \, d^4x \\ &= \int N^{\mu\nu}(-g_{\mu\alpha} b^\alpha_{,\nu} - g_{\nu\alpha} b^\alpha_{,\mu} - g_{\mu\nu,\sigma} b^\sigma) \sqrt{\ } \, d^4x \\ &= \int [2(N_\alpha{}^\nu \sqrt{\ })_{,\nu} - g_{\mu\nu,\alpha} N^{\mu\nu} \sqrt{\ }] b^\alpha \, d^4x \\ &= 2 \int N_\alpha{}^\nu{}_{:\nu} b^\alpha \sqrt{\ } \, d^4x \end{aligned}$$

from the theorem expressed by (21.4), which is valid for any symmetrical two-index tensor. The invariance property of I' requires that it shall be unaltered under this variation, for all b^α. Hence $N_\alpha{}^\nu{}_{:\nu} = 0$.

On account of this relation, the field equations (30.4), (30.5) are not all independent.

31. The pseudo-energy tensor of the gravitational field

Define the quantity $t_\mu{}^\nu$ by

$$t_\mu{}^\nu \surd = \frac{\partial \mathscr{L}}{\partial g_{\alpha\beta,\nu}} g_{\alpha\beta,\mu} - g_\mu^\nu \mathscr{L}. \tag{31.1}$$

We then have

$$(t_\mu{}^\nu \surd)_{,\nu} = \left(\frac{\partial \mathscr{L}}{\partial g_{\alpha\beta,\nu}}\right)_{,\nu} g_{\alpha\beta,\mu} + \frac{\partial \mathscr{L}}{\partial g_{\alpha\beta,\nu}} g_{\alpha\beta,\mu\nu} - \mathscr{L}_{,\mu}.$$

Now

$$\mathscr{L}_{,\mu} = \frac{\partial \mathscr{L}}{\partial g_{\alpha\beta}} g_{\alpha\beta,\mu} + \frac{\partial \mathscr{L}}{\partial g_{\alpha\beta,\nu}} g_{\alpha\beta,\nu\mu},$$

so

$$\begin{aligned}(t_\mu{}^\nu \surd)_{,\nu} &= \left[\left(\frac{\partial \mathscr{L}}{\partial g_{\alpha\beta,\nu}}\right)_{,\nu} - \frac{\partial \mathscr{L}}{\partial g_{\alpha\beta}}\right] g_{\alpha\beta,\mu} \\ &= (16\pi)^{-1}(R^{\alpha\beta} - \tfrac{1}{2}g^{\alpha\beta}R) g_{\alpha\beta,\mu}\surd\end{aligned}$$

from (30.2). With the help of the field equations (24.6) we now get

$$(t_\mu{}^\nu \surd)_{,\nu} = -\tfrac{1}{2}Y^{\alpha\beta} g_{\alpha\beta,\mu}\surd,$$

so from (21.4) and $Y_\mu{}^\nu{}_{:\nu} = 0$, we get

$$[(t_\mu{}^\nu + Y_\mu{}^\nu)\surd]_{,\nu} = 0. \tag{31.2}$$

We have here a conservation law, and it is natural to consider the conserved density $(t_\mu{}^\nu + Y_\mu{}^\nu)\surd$ as the density of energy and momentum. We have already had $Y_\mu{}^\nu$ as the energy and momentum of the fields other than the gravitational field, so $t_\mu{}^\nu$ represents the energy and momentum of the gravitational field. But *it is not a tensor*. The equation (31.1) that defines it may be written

$$t_\mu{}^\nu = \frac{\partial L}{\partial g_{\alpha\beta,\nu}} g_{\alpha\beta,\mu} - g_\mu^\nu L; \tag{31.3}$$

but L is not a scalar, because we had to transform the scalar R, which was originally used to get the gravitational action, in order to remove the second derivatives from it. Thus $t_\mu{}^\nu$ cannot be a tensor. It is called a pseudo-tensor.

It is not possible to obtain an expression for the energy of the gravitational field satisfying both the conditions: (i) when added to other forms of energy the total energy is conserved, and (ii) the energy within a definite (three-dimensional) region at a certain time is independent of the coordinate system. Thus, in general, *gravitational energy cannot be localized.* The best we can do is to use the pseudo-tensor, which satisfies condition (i) but not (ii). It gives us approximate information about gravitational energy, which in some special cases can be accurate.

We may form the integral

$$\int (t_\mu{}^0 + Y_\mu{}^0)\sqrt{\;}\, dx^1\, dx^2\, dx^3 \tag{31.4}$$

over a large three-dimensional volume enclosing some physical system at a certain time. As the volume tends to infinity, we may suppose the integral to give the total energy and momentum, provided: (a) it converges and (b) the flux through the surface of the large volume tends to zero. The equation (31.2) then shows that the integral (31.4) taken at one time $x^0 = a$ equals its value at another time $x^0 = b$. Furthermore, the integral must be independent of the coordinate system, since we could change the coordinates at $x^0 = b$ without changing them at $x^0 = a$. We thus have definite expressions for the total energy and momentum, which are conserved.

The conditions (a) and (b), which are needed for conservation of total energy and momentum, do not often apply in practical cases. They would apply if space were static outside a definite tubular region in four dimensions. This could be so if we had some masses which start to move at a certain time, so that the motion creates a disturbance which travels outward with the velocity of light. For the usual planetary system the motion will have been going on since the infinite past and the conditions do not apply. A special treatment is needed to discuss the energy of the gravitational waves, and this will be given in Section 33.

32. Explicit expression for the pseudo-tensor

The formula (31.1) for defining $t_\mu{}^\nu$ is of the form

$$t_\mu{}^\nu \surd = \frac{\partial \mathscr{L}}{\partial q_{n,\nu}} q_{n,\mu} - g_\mu^\nu \mathscr{L}, \tag{32.1}$$

where the q_n $(n = 1, 2, \ldots, 10)$ are the ten $g_{\mu\nu}$ and a summation over all n is implied. We could equally well write it

$$t_\mu{}^\nu \surd = \frac{\partial \mathscr{L}}{\partial Q_{m,\nu}} Q_{m,\mu} - g_\mu^\nu \mathscr{L}, \tag{32.2}$$

where the Q_m are any ten independent functions of the q_n. To prove this, note that

$$Q_{m,\sigma} = \frac{\partial Q_m}{\partial q_n} q_{n,\sigma}.$$

Hence

$$\frac{\partial \mathscr{L}}{\partial q_{n,\nu}} = \frac{\partial \mathscr{L}}{\partial Q_{m,\sigma}} \frac{\partial Q_{m,\sigma}}{\partial q_{n,\nu}} = \frac{\partial \mathscr{L}}{\partial Q_{m,\sigma}} \frac{\partial Q_m}{\partial q_n} g_\sigma^\nu$$
$$= \frac{\partial \mathscr{L}}{\partial Q_{m,\nu}} \frac{\partial Q_m}{\partial q_n}.$$

Thus

$$\frac{\partial \mathscr{L}}{\partial q_{n,\nu}} q_{n,\mu} = \frac{\partial \mathscr{L}}{\partial Q_{m,\nu}} \frac{\partial Q_m}{\partial q_n} q_{n,\mu} = \frac{\partial \mathscr{L}}{\partial Q_{m,\nu}} Q_{m,\mu}.$$

The equality of (32.1) and (32.2) follows.

To deduce an explicit expression for $t_\mu{}^\nu$ it is convenient to work with (32.2) and to take the Q_m to be the quantities $g^{\mu\nu}\surd$. We can now use formula (26.7), which gives (bringing in the coefficient 16π),

$$16\pi\, \delta\mathscr{L} = (\Gamma^\nu_{\alpha\beta} - g^\nu_\beta \Gamma^\sigma_{\alpha\sigma})\, \delta(g^{\alpha\beta}\surd)_{,\nu} + (\text{some coeft})\, \delta(g^{\mu\nu}\surd),$$

and hence

$$16\pi t_\mu{}^\nu \surd = (\Gamma^\nu_{\alpha\beta} - g^\nu_\beta \Gamma^\sigma_{\alpha\sigma})(g^{\alpha\beta}\surd)_{,\mu} - g^\nu_\mu \mathscr{L}. \tag{32.3}$$

33. Gravitational waves

Let us consider a region of empty space where the gravitational field is weak and the $g_{\mu\nu}$ are approximately constant. We then have equation (16.4) or

$$g^{\mu\nu}(g_{\mu\nu,\rho\sigma} - g_{\mu\rho,\nu\sigma} - g_{\mu\sigma,\nu\rho} + g_{\rho\sigma,\mu\nu}) = 0. \tag{33.1}$$

Let us take harmonic coordinates. The condition (22.2) gives, with the suffix λ lowered,

$$g^{\mu\nu}(g_{\rho\mu,\nu} - \tfrac{1}{2}g_{\mu\nu,\rho}) = 0. \tag{33.2}$$

Differentiate this equation with respect to x^σ and neglect second-order terms. The result is

$$g^{\mu\nu}(g_{\mu\rho,\nu\sigma} - \tfrac{1}{2}g_{\mu\nu,\rho\sigma}) = 0. \tag{33.3}$$

Interchange ρ and σ:

$$g^{\mu\nu}(g_{\mu\sigma,\nu\rho} - \tfrac{1}{2}g_{\mu\nu,\rho\sigma}) = 0. \tag{33.4}$$

Add (33.1), (33.3), and (33.4). We get

$$g^{\mu\nu}g_{\rho\sigma,\mu\nu} = 0.$$

Thus each $g_{\rho\sigma}$ satisfies the d'Alembert equation and its solution will consist of waves traveling with the velocity of light. They are gravitational waves.

Let us consider the energy of these waves. Owing to the pseudo-tensor not being a real tensor, we do not get, in general, a clear result independent of the coordinate system. But there is one special case in which we do get a clear result; namely, when the waves are all moving in the same direction.

If the waves are all moving in the direction x^3, we can choose our coordinate system so that the $g_{\mu\nu}$ are functions of only the one variable $x^0 - x^3$. Let us take the more general case in which the $g_{\mu\nu}$ are all functions of the single variable $l_\sigma x^\sigma$, the l_σ being constants satisfying $g^{\rho\sigma}l_\rho l_\sigma = 0$, with neglect of the variable part of the $g^{\rho\sigma}$. We then have

$$g_{\mu\nu,\sigma} = u_{\mu\nu}l_\sigma, \tag{33.5}$$

where $u_{\mu\nu}$ is the derivative of the function $g_{\mu\nu}$ of $l_\sigma x^\sigma$. Of course, $u_{\mu\nu} = u_{\nu\mu}$. The harmonic condition (33.2) gives

$$g^{\mu\nu}u_{\mu\rho}l_\nu = \tfrac{1}{2}g^{\mu\nu}u_{\mu\nu}l_\rho = \tfrac{1}{2}ul_\rho,$$

with $u = u^\mu_\mu$. We may write this as

$$u^\nu_\rho l_\nu = \tfrac{1}{2}ul_\rho \tag{33.6}$$

or as

$$(u^{\mu\nu} - \tfrac{1}{2}g^{\mu\nu}u)l_\nu = 0. \tag{33.7}$$

We have from (33.5)

$$\Gamma^\rho_{\mu\sigma} = \tfrac{1}{2}(u^\rho_\mu l_\sigma + u^\rho_\sigma l_\mu - u_{\mu\sigma}l^\rho).$$

The expression (26.3) for L reduces, with harmonic coordinates, to

$$\begin{aligned} L &= -g^{\mu\nu}\Gamma^\rho_{\mu\sigma}\Gamma^\sigma_{\nu\rho} \\ &= -\tfrac{1}{4}g^{\mu\nu}(u^\rho_\mu l_\sigma + u^\rho_\sigma l_\mu - u_{\mu\sigma}l^\rho)(u^\sigma_\nu l_\rho + u^\sigma_\rho l_\nu - u_{\nu\rho}l^\sigma). \end{aligned}$$

This gives nine terms when multiplied up, but we can easily see that every one of them vanishes, on account of (33.6) and $l_\sigma l^\sigma = 0$. Thus the action density vanishes. There is a corresponding result for the electromagnetic field, for which the action density also vanishes in the case of waves moving only in one direction.

We must now evaluate the pseudo-tensor (32.3). We have

$$\begin{aligned} g^{\alpha\beta}{}_{,\mu} &= -g^{\alpha\rho}g^{\beta\sigma}g_{\rho\sigma,\mu} = -u^{\alpha\beta}l_\mu, \\ \surd_{,\mu} &= \tfrac{1}{2}\surd g^{\alpha\beta}g_{\alpha\beta,\mu} = \tfrac{1}{2}\surd ul_\mu, \end{aligned} \tag{33.8}$$

so

$$(g^{\alpha\beta}\surd)_{,\mu} = -(u^{\alpha\beta} - \tfrac{1}{2}g^{\alpha\beta}u)\surd l_\mu.$$

Hence

$$\begin{aligned} \Gamma^\sigma_{\alpha\sigma}(g^{\alpha\beta}\surd)_{,\mu} &= \surd_{,\alpha}(-u^{\alpha\beta} + \tfrac{1}{2}g^{\alpha\beta}u)l_\mu \\ &= 0, \end{aligned}$$

from (33.8) and (33.7). We are left with

$$\begin{aligned} 16\pi t_\mu{}^\nu &= -\Gamma^\nu_{\alpha\beta}(u^{\alpha\beta} - \tfrac{1}{2}g^{\alpha\beta}u)l_\mu \\ &= -\tfrac{1}{2}(u^\nu_\alpha l_\beta + u^\nu_\beta l_\alpha - u_{\alpha\beta}l^\nu)(u^{\alpha\beta} - \tfrac{1}{2}g^{\alpha\beta}u)l_\mu \\ &= \tfrac{1}{2}(u_{\alpha\beta}u^{\alpha\beta} - \tfrac{1}{2}u^2)l_\mu l^\nu. \end{aligned} \tag{33.9}$$

We have a result for $t_\mu{}^\nu$ that looks like a tensor. This means that $t_\mu{}^\nu$ transforms like a tensor under those transformations that preserve the character of the field of consisting only of waves moving in the direction l_σ, so that the $g_{\mu\nu}$ remain functions of the single variable $l_\sigma x^\sigma$. Such transformations must consist only in the introduction of coordinate waves moving in the direction l_σ, of the form

$$x^{\mu'} = x^\mu + b^\mu,$$

where b_μ is a function only of $l_\sigma x^\sigma$. With the restriction that we have waves moving only in one direction, gravitational energy can be localized.

34. The polarization of gravitational waves

To understand the physical significance of (33.9), let us go back to the case of waves moving in the direction x^3, so that $l_0 = 1, l_1 = l_2 = 0, l_3 = -1$, and use coordinates approximating to those of special relativity. The harmonic conditions (33.6) now give

$$u_{00} + u_{03} = \tfrac{1}{2}u,$$
$$u_{10} + u_{13} = 0,$$
$$u_{20} + u_{23} = 0,$$
$$u_{30} + u_{33} = -\tfrac{1}{2}u.$$

Thus

$$u_{00} - u_{33} = u = u_{00} - u_{11} - u_{22} - u_{33},$$

so

$$u_{11} + u_{22} = 0. \tag{34.1}$$

Also

$$2u_{03} = -(u_{00} + u_{33}).$$

We now get

$$\begin{aligned} u_{\alpha\beta}u^{\alpha\beta} - \tfrac{1}{2}u^2 &= u_{00}{}^2 + u_{11}{}^2 + u_{22}{}^2 + u_{33}{}^2 - 2u_{01}{}^2 - 2u_{02}{}^2 \\ &\quad - 2u_{03}{}^2 + 2u_{12}{}^2 + 2u_{23}{}^2 + 2u_{31}{}^2 - \tfrac{1}{2}(u_{00} - u_{33})^2 \\ &= u_{11}{}^2 + u_{22}{}^2 + 2u_{12}{}^2 \\ &= \tfrac{1}{2}(u_{11} - u_{22})^2 + 2u_{12}{}^2, \end{aligned}$$

from (34.1). Thus

$$16\pi t_0{}^0 = \tfrac{1}{4}(u_{11} - u_{22})^2 + u_{12}{}^2 \tag{34.2}$$

and

$$t_0{}^3 = t_0{}^0.$$

We see that the energy density is positive definite and the energy flows in the direction x^3 with the velocity of light.

To discuss the polarization of the waves, we introduce the infinitesimal rotation operator R in the plane x^1x^2. Applied to any vector A_1, A_2, it has the effect

$$RA_1 = A_2, \qquad RA_2 = -A_1.$$

Thus

$$R^2A_1 = -A_1,$$

so iR has the eigenvalues ± 1 when applied to a vector.

Applied to $u_{\alpha\beta}$, it has the effect

$$Ru_{11} = u_{21} + u_{12} = 2u_{12},$$

$$Ru_{12} = u_{22} - u_{11},$$

$$Ru_{22} = -u_{12} - u_{21} = -2u_{12}.$$

So

$$R(u_{11} + u_{22}) = 0$$

and

$$R(u_{11} - u_{22}) = 4u_{12}$$

$$R^2(u_{11} - u_{22}) = -4(u_{11} - u_{22}).$$

Thus $u_{11} + u_{22}$ is invariant, while iR has the eigenvalues ± 2 when applied to $u_{11} - u_{22}$ or u_{12}. The components of $u_{\alpha\beta}$ that contribute to the energy (34.2) thus correspond to spin 2.

35. The cosmological term

Einstein has considered generalizing his field equations for empty space to

$$R_{\mu\nu} = \lambda g_{\mu\nu}, \tag{35.1}$$

where λ is a constant. This is a tensor equation, so it is permissible as a law of nature.

We get good agreement with observation for the solar system without this term, and therefore if we do introduce it we must take λ to be small enough not to disturb the agreement. Since $R_{\mu\nu}$ contains second derivatives of the $g_{\mu\nu}$, λ must have the dimensions (distance)$^{-2}$. For λ to be small this distance must be very large. It is a cosmological distance, of the order of the radius of the universe.

The extra term is important for cosmological theories, but has a negligible effect on the physics of nearby objects. To take it into account in the field theory, we merely have to add an extra term to the Lagrangian; namely,

$$I_c = c\int \sqrt{\;}\, d^4x,$$

with c a suitable constant.

We have from (26.10)

$$\delta I_c = c\int \tfrac{1}{2}g^{\mu\nu}\,\delta g_{\mu\nu}\sqrt{\;}\, d^4x.$$

Thus the action principle

$$\delta(I_g + I_c) = 0$$

gives

$$16\pi(R^{\mu\nu} - \tfrac{1}{2}g^{\mu\nu}R) + \tfrac{1}{2}cg^{\mu\nu} = 0. \tag{35.2}$$

The equation (35.1) gives

$$R = 4\lambda,$$

and hence

$$R^{\mu\nu} - \tfrac{1}{2}g_{\mu\nu}R = -\lambda g_{\mu\nu}.$$

This agrees with (35.2), provided we take

$$c = 32\pi\lambda.$$

For the gravitational field interacting with any other fields, we merely have to include the term I_c in the action and we will get the correct field equations with Einstein's cosmological term.

Index